FEATHERED DINOSAURS

FEATHERED DINOSAURS

THE ORIGIN OF BIRDS

JOHN LONG AND PETER SCHOUTEN
FOREWORD BY LUIS M. CHIAPPE

CSIRO
PUBLISHING

OXFORD
UNIVERSITY PRESS

OXFORD
UNIVERSITY PRESS

Oxford University Press, Inc., publishes works that further
Oxford University's objective of excellence
in research, scholarship, and education.

Oxford New York
Auckland Cape Town Dar es Salaam Hong Kong Karachi
Kuala Lumpur Madrid Melbourne Mexico City Nairobi
New Delhi Shanghai Taipei Toronto

With offices in
Argentina Austria Brazil Chile Czech Republic France Greece
Guatemala Hungary Italy Japan Poland Portugal Singapore
South Korea Switzerland Thailand Turkey Ukraine Vietnam

Published in the United States by
Oxford University Press, Inc.
198 Madison Avenue, New York, New York 10016

www.oup.com

Originally published by CSIRO Publishing, Australia

Oxford is a registered trademark of Oxford University Press

Library of Congress Cataloging-in-Publication Data
Long, John A., 1957–
Feathered dinosaurs : the origin of birds / John Long and Peter Schouten ;
foreword by Luis M. Chiappe.
      p. cm.
Includes bibliographical references and index.
ISBN 978-0-19-537266-3 (cloth : alk. paper) 1. Dinosaurs. 2. Birds, Fossil. 3. Birds—Origin.
I. Schouten, Peter. II. Title.
QE862.D5L582 2008
567.9—dc22   2008001232

1 3 5 7 9 8 6 4 2

Set in 9/14 Optima
Cover and text design by James Kelly
Printed in Singapore by Imago

FOREWORD

I therefore believe a unification of art and science to be the highest goal, and especially
palaeontology will always make a poor and meagre expression without the aid of art.

GERHARD HEILMANN, 1916

The marriage of art and science has a tradition that stretches back to Leonardo da Vinci and which through the genius of twentieth century artists such as Charles R. Knight, Gerhard Heilmann and Zdeněk Burian, has become an integral aspect of modern palaeontology. In *Feathered Dinosaurs: The Origin of Birds*, palaeontologist John Long and artist Peter Schouten work together to provide us with an unprecedented visual record of one of the most significant breakthroughs in the history of vertebrate palaeontology – the discovery that a diversity of predatory dinosaurs were cloaked with feathers, perhaps just as colorful and fanciful as those of their living relatives.

The idea that the ancestry of birds can be traced back to carnivorous dinosaurs called theropods is not new. Nearly 150 years ago, British anatomist Thomas Henry Huxley promulgated that there was nothing wild or illegitimate about this evolutionary connection. In the ensuing years, a myriad of skeletal features supporting the dinosaurian origin of birds was discovered in the fossils of large and small theropods – hollowed bones, elongated forelimbs, large breastbones, and even wishbones are some of the features typical of the avian skeleton that have by now been found among these dinosaurs. Discoveries of the last few decades have also added new lines of evidence in support of the same evolutionary relationship. The shape of the eggs and the microscopic structure of the eggshell of theropods were shown to be similar to that of birds as well as the pattern of egg-laying of all these animals was found to be comparable. Snapshots of ancient behaviour provided by a handful of exceptional fossils brought even further support to the hypothesis that birds evolved from theropod dinosaurs. These discoveries proved that some of these dinosaurs brooded their offspring in a fashion similar to that of birds and that they adopted resting poses that would be familiar to any birder. More specific fields of research have made their own empirical contributions in support of the theropod legacy of birds. Studies of dinosaurian growth rates, based on details preserved in the fossilised tissue of their bones, have documented that these once-believed sluggish animals actually grew at speeds comparable to

> Snapshots of ancient behaviour provided by a handful of exceptional fossils brought even further support to the hypothesis that birds evolved from theropod dinosaurs.

many living birds. And even studies correlating the sizes of bone-cells and genomes (the entire genetic material of an organism) have revealed that the mighty *T. rex* and its fearsome kin had the small genome typical of modern birds. Yet, despite the multiplicity of this extensive body of evidence, nothing has cemented the dinosaurian pedigree of birds more than the realisation that true feathers covered the bodies of a variety of animals which, by virtue of their colossal size, might, and lasting dominance have become lore of a bygone world.

The enormous significance of these fossils notwithstanding, the documented existence of feathers in dinosaurs has thus far been limited to a dozen or so species, all of them circumscribed to East Asia. But because these fossils span a large portion of the family tree of theropods and display a great diversity of sizes, appearances, and lifestyles, this handful of feathered dinosaurs hints at a much larger and yet undocumented diversity. In *Feathered Dinosaurs* this modest roster of fossils is expanded to include a much larger menagerie, whose plumage is justly inferred on the basis of their kinship to the few dinosaurs known of have had feathering.

In the last few decades, our understanding of the origin and subsequent evolutionary diversification of birds has advanced at an unparalleled pace. Not only has the ancestry of birds been greatly clarified but the large fossil gap that separated living birds from their dinosaurian predecessors has been filled with numerous findings. These discoveries have established beyond any reasonable doubt that birds evolved from carnivorous dinosaurs gifted with the plumages and exuberant behaviors that we see today in their descendants. The prehistoric re-enactments of *Feathered Dinosaurs* – radiant portraits of dinosaur life immersed in dramatic scenographies – remind us once again of the harmonious communion between science and art, and the importance of visualisation in the study of ancient life.

Luis M. Chiappe
Director, The Dinosaur Institute
Natural History Museum of Los Angeles County, Los Angeles

CONTENTS

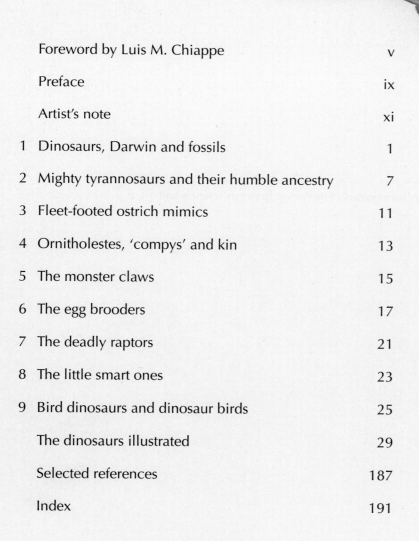

PREFACE

Sleep softly... eagle forgotten... under the stone.
Time has its way with you there, and the clay has its own[1]

The first two lines of Illinois poet Nicholas Vachel Lindsay's poem conjure up a haunting vision of a once mighty and noble bird, now gone and buried by the stone and clay of a past age. Eagles, hawks and falcons are today collectively known as the birds of prey, the 'raptors'. Yet in modern culture the word raptor has recently taken on a new meaning since the increased interest in dinosaurs brought about by films such as the *Jurassic Park* series by Steven Spielberg. Many people today associate the word 'raptor' with the cunning, swift, predatory dinosaurs that hunted in packs and could seemingly solve complex problems in order to get to their prey, deriving the term from the diminutive of the name of the most infamous beast, *Velociraptor*.

Velociraptor, a small carnivorous dinosaur that lived almost 80 million years ago in Mongolia, is a nightmare from an ancient bygone world. Like Lindsay's *The Eagle That is Forgotten* it too is a raptor entombed by the clays and stone of a world past. However, new studies on the fossil remains of dinosaurs known as theropods – meaning 'beast-footed' ones – mark an epiphany in our understanding of evolution. For now we are discovering that dinosaurs like *Velociraptor* are the ancestors of the modern birds we know today. However, the most startling revelation about *Velociraptor* and its kin is that many of them are now known to have possessed feathers. This fact has made us think again not only about their evolutionary transition to birds, but also about how they might have used feathers in their daily life. Did they use feathers in complex mating rituals? Did they use them to brood their young? Or did feathers act primarily as a stepping-stone in the evolution of flight? We know from fossil evidence that some of these scenarios, and possibly all of them, were true.

In this book we look at the various species of dinosaurs that belong to major groups, in which we know at least some of the members possessed feathers. From this observation scientists infer that, as a general rule, most dinosaurs in these groups would also have had feathers for at least some part of their lives, either just as juveniles or retaining them throughout their lives.

> ... the most startling revelation about *Velociraptor* and its kin is that many of them are now known to have possessed feathers.

---

1 N. V. Lindsay (1902), *The Eagle That is Forgotten*, written as a tribute to the memory of US politician John Peter Altgeld.

We show the evolutionary march from primitive, feathered dinosaurs through to the first true birds. We present the salient facts about these dinosaurs and discuss the extent of information that can be deduced from their fossil remains. The artist has exercised his extensive knowledge of the lifestyles of modern birds and mammals to marry the scientific data to a vision of how each creature might have looked and behaved. We make no apologies about using our imaginations to the fullest in presenting what we hope are the most accurate life-like portraits of these exotic bygone creatures. Our reconstructions also serve to dramatically depict the various transitions between the carnivorous dinosaurs and modern birds, dispelling any doubt about the veracity of evolutionary processes.

I would like to thank Dr Luis Chiappe, Dr Thomas Rich, Brian Choo and Heather Robinson Long for providing helpful discussion on dinosaur and early bird evolution during the writing of this book and for reading the draft manuscript. My thanks also to Dr Zhonghe Zhou for information about early Chinese fossil birds.

John Long
Head of Sciences, Museum Victoria, Melbourne

ARTIST'S NOTE

From earliest childhood I have been fascinated by the wondrous world of the dinosaur. As an adult, this interest has drawn me down many a musty museum corridor and into many a dusty drawer as I began my early career in palaeontological reconstruction – one which subsequently developed into my current vocation as a wildlife artist. When new fossil evidence confronted our previously held notions about dinosaurs, an opportunity was presented to me to unite the disciplines of my past with those of my present – and it was an opportunity the fascinated child still within me could not resist.

The art of reconstructing creatures from ages long past is a difficult one. Tempting though it would be to let the imagination run riot, one must instead undertake painstaking research to ensure that the finished painting is the most accurate depiction possible. To determine the size and structure of the animal or distinctive features such as teeth etc, the fossil record must be closely consulted. The fossil record, too, has been useful in revealing evidence of habitat, such as the presence of pollen or other plant material, and where possible a suggestion of that habitat has been incorporated into the paintings. Unfortunately, no matter how comprehensive the fossil record is, it only rarely provides clues to assist us in determining a plausible outward appearance of these creatures long gone.

We must instead look elsewhere for the information that will ensure a credible depiction upon the canvas. In particular, there is one avenue that provides a wealth of information – the study of evolution. The popular idea that a catastrophic asteroid impact, some 65 million years ago, completely obliterated the world of the dinosaur is one that is not entirely correct. Although around 75 per cent of life on the planet was destroyed, it should be remembered that many animals survived and continued to evolve, and it is the descendants of this ancient life that provide the information we seek. In the case of the dinosaurs, the study of the behaviour and morphology of their closest living relatives – birds – has proven invaluable.

For the paintings in this book, I have drawn inspiration from existing creatures when determining features such as skin colouring and covering, eye size, iris colouration and so forth. In my reconstructions I was guided by the three fundamentals of the animal kingdom – the need to eat, the need to avoid being eaten and the desire to procreate. Depending on habitat, these fundamentals can have an extraordinary effect upon a creature's outward appearance, and thus I have given considerable consideration to the habitats and the locale in which the fossils were found.

Accompanying each painting throughout this book is a small aside from me to explain the reasons why I have reconstructed the animal in the manner that I have. Where a creature is known to the fossil record only by a fragment – say the skull – I have chosen to reconstruct only that part of the animal and have left the rest to your imagination. My imagination has come to the fore in the compositions, where it is more dramatic (and fun) to reconstruct a dinosaur tearing flesh than say, sleeping. Having said that, I have tried to illustrate an aspect of behaviour, morphology or a physical trait unique to that species and the groups to which they belong.

The paintings that follow are not a comprehensive depiction of those bird-like dinosaurs with feathers – the Coelurosauria. The animals shown here are only those for which sufficient material exists to provide a reasonably accurate reconstruction. I hope you receive as much enjoyment looking at these paintings as I had painting them.

Peter Schouten
April 2007

1

DINOSAURS, DARWIN AND FOSSILS

Of the many strange forms of ancient animal life brought to light by the labours of geologists and collectors in many parts of the world, perhaps those of the Dinosaurian order are the most wonderful. Not only in size, but in strangeness and variety, they may be said almost to stand alone. They were indeed the veritable dragons of old time, and seem to come nearer than any other antediluvian animals to the monsters of fairyland, whose acquaintance we all made in the nursery.[2]

Dinosaurs were the supreme rulers of our planet for around 160 million years. It was during their reign that the modern kinds of mammals and the first birds appeared and diversified. Although first perceived and depicted artistically as somewhat savage beasts of low intelligence – some having 'a brain the size of a walnut' – dinosaurs were considered by nineteenth century scientists as being not much more than giant lizards.

The Dinosauria (meaning 'terrible lizard') were first recognised as a distinct class of reptiles by the famous British anatomist Sir Richard Owen in 1842, based upon a few scant fossil bones. These demonstrated that these reptilian monsters walked with almost vertical leg postures, like we humans and other mammals do, and so he perceived dinosaurs as being quite distinctly separate from the more lowly, sprawling kinds of reptile.

The modern study of dinosaurs involves the disciplines of palaeontology, comparative anatomy, developmental biology and genetics. It also incorporates refined CT-scanning, histological and biochemical studies, to analyse the old fossil bones in detail as never before. In recent years such research has presented fresh new interpretations on how dinosaurs grew, lived, laid their eggs, brooded their young, digested their food, breathed, retained body heat, and even how some made elaborate nests. But it is the breathtaking new fossil evidence from sites in Liaoning Province in north-eastern China, uncovered only during the past decade or so, which demonstrates that many dinosaurs had feathers, or feathery hair-like coverings, on their bodies.

... breathtaking new fossil evidence from sites in Liaoning Province in north-eastern China ... demonstrates that many dinosaurs had feathers, or feathery hair-like coverings, on their bodies.

---

2 Rev H. N. Hutchinson (1894), *Creatures of Other Days*, p. 117.

Such finds, combined with what we already knew of the skeletal similarities shared by birds and predatory dinosaurs, leave little doubt about the relationship between feathered dinosaurs and today's birds.

We know of dinosaurs and early birds from their fossils. Fossils are the remains of past life on the planet. They can be bones, shells, wood, trackways, burrows, trails, insects caught in amber, or even fossilised tissues replaced by microbial minerals to replicate the soft organs of a once-living creature. Fossils are our principal way of looking back into the past to see what life was doing eons ago, how environments were developing with shifts in climate, and how the interminably slow drift of the continents shaped the modern face of planet Earth.

The fossil record, as held in all the world's major museums, government organisations, universities and private collections, now tallies close to a billion fossil specimens. Most come from a sedimentary rock layer, or stratum, that depicts the chronological sequence of diversification of life known as evolution. If just one of those billion specimens was irrefutably demonstrated to be out of the predicted context, then there would indeed be grounds to debate evolution. However, apart from obvious cases of fossils reworked from one older deposit into another younger one, the world's sum collection of fossils testifies to the absolute veracity and reliability of the theory of evolution. Remarkable new discoveries from both old and new sites around the world have been slowly filling in many of the gaps in the evolutionary record of the back-boned animals, or vertebrates – those creatures having 'vertebrae'.

Today we know of intermediate forms of fishes with limbs and skulls indistinguishable from those of early amphibians, whose rising from the waters about 360 million years ago was accompanied by the development of terrestrial adaptations in their skeleton necessary to support their life on land. We see every possible step in the transition from early amphibians to reptiles and from reptiles to mammals, seen principally by the simplification of the complex reptilian jaw to the simple mammalian mandible having just one bone, the dentary. At the same time, the bones that attached once to the lower jaw of reptiles were not lost. Instead, they shifted in towards the ear and became used to transfer sound. These are now the three standard bones we use for hearing: the incus, malleus and stapes.

Reptiles underwent a great radiation in the Mesozoic Era (250 to 65 million years ago) culminating in what could be claimed as the grandest of all creatures, the dinosaurs. At the same time as the first dinosaurs appeared, we see also the advent of early mammals and flying pterosaurs. All three groups flourished during the three periods of the Mesozoic Era, the Triassic, Jurassic and Cretaceous, but the dinosaurs dominated the main niches on land, whilst the reptilian pterosaurs ruled the skies. Mammals, which appeared about 220 million years ago, and birds, which appeared about 150 million years ago, were both very much subordinate parts of the Mesozoic ecosystem. Yet despite the amazing record of fossils we have in our museums, fossils were not the foundation for ideas that formed our modern concept of evolution. Instead these concepts came from observations of the modern biological world, principally by Charles Darwin but also from the work of Alfred Russel Wallace.

The small carnivorous dromaeosaur, *Atrociraptor*, is known from only one fossil find, comprising parts of its jaws and teeth.

In Charles Darwin's days there were actually very few fossils that illustrated the concept of evolution or showed a transition of form between any of the major groups of animals. The origin of birds was an unsolved question when *The Origin of Species* was first published in 1859. The oldest primitive fossil bird from the age of dinosaurs, one showing a suite of reptilian features, was not discovered until a few years later. This fossil was uncovered in southern Germany in 1861, although a single feather heralding the fact that birds were around at that time had been found a year earlier from the same site. The 1861 find was of a complete fossil bird identified as *Archaeopteryx lithographica*, a name given by the German scientist von Meyer, based on the earlier feather fossil. It was a magnificent specimen showing clearly a creature with wings adorned by elaborate flight feathers. The British Museum of Natural History in London absolutely had to have it at all costs so purchased the specimen from the finder for the inordinate amount of £700, paying from funds set aside over two budgetary years.

The ardent and authoritative Sir Richard Owen of the British Museum was a prolific anatomist with a weight of scholarly tomes behind him. He was also a firm believer in God's creation, which implied there should be no intermediary forms between the major kinds of created organisms. But Owen did accept that species changed. His work on the evolution of horses demonstrated that he accepted the transmutation of species, but unlike Darwin he saw evolution as the gradual unfolding of a divine plan.

Owen wrote the first detailed account of *Archaeopteryx*, published in 1863, but concluded that it was merely a bird that showed some adaptations to living like a reptile. His work implied it was simply a bird that had secondarily developed some reptilian features. A year after Owen's description of *Archaeopteryx*, William Kitchen Parker, a medical doctor and foremost expert on the anatomy of living birds (who later wrote the section on birds for *Encyclopaedia Britannica*), gave another short account of the *Archaeopteryx* fossil. In summarising he made poetic comparisons, bringing out his unabashed emotional feelings concerning birds and reptiles.

The discovery in 1860 of a single fossil feather from *Archaeopteryx* provided the first clue as to the origin of birds.

> *There is a curious blending of the characters of the various reptilian groups in the Birds; there has been no exclusive adoption of the mode of structure of any one scaly type by these feathered vertebrates; those reptilian qualities and excellencies which are best and highest have become theirs; how much more! This exaltation of the 'Sauropsidan' or oviparous type by the substitution of feathers for scales, wings for paws, warm blood for cold, intelligence for stupidity, and what is lovely instead of loathsomeness, – this sudden glorification of the vertebrate form is one of the great wonders of Nature.*[3]

In 1867 Thomas Henry Huxley, often known as 'Darwin's bulldog' for his fervent support of the evolutionary theory, was the first scientist to show that *Archaeopteryx* had many other reptilian features in its skeleton, features which had been overlooked by Owen. Yet, like the story of the 'Emperor's New Clothes', some scientists could not be swayed to see *Archaeopteryx* for its true value as 'a missing link' joining the modern birds to the reptiles through Darwin's theory of evolution.

---

3   W. K. Parker (1864), *Geological Magazine* 1, pp. 56–57.

Huxley's work was indeed controversial. Owen had erected the Class Dinosauria in 1842 and was at the time deemed to be the leading expert on this strange extinct group of giant reptiles. Huxley had shown that *Archaeopteryx* – and indeed any modern bird – has more in common with the predatory theropod dinosaurs than any other reptile. Other scientists of the day sided with Owen and saw *Archaeopteryx* as just a bird, thereby maintaining the *status quo* of the Church's view on evolution, and pooh-poohing any notion of a 'missing link' in the fossil record.

In the early half of the twentieth century, other discoveries of dinosaurs and their reptilian ancestors emerged from Triassic-age deposits. In 1869 the rambunctious American dinosaur hunter Edward Drinker Cope coined the term 'Archosauria' (meaning 'ruling reptiles') for a group of living and extinct reptiles. In 1945 leading American palaeontologist Alfred Sherwood Romer refined this term to include dinosaurs and all reptiles with similarities to dinosaurs, such as crocodiles, the flying pterosaurs and groups of primitive Triassic forms. Archosaurs are an interesting group of dinosaur-like reptiles united by having several unique features, the most notable being the skull with a large hole in front of the eye.

Troodon had large wings that could fold back like a bird's.

A mixed bag of odd Triassic reptiles formerly called the 'thecodonts' (meaning 'tooth-socketed' ones) were once generally accepted as the most likely source group for dinosaurian origins. Today though, with refined techniques of classification, we do not recognise thecodonts as one coherent group but as a series of lineages of archosaurian reptiles that may not be directly interrelated to each other. In addition to having a skull with large holes in it for the jaw musculature to flex, as in later dinosaurs, the foot structure of some of these reptiles shows adaptations to having a variable gait, that is with a semi-erect posture of the legs. Though not quite a dinosaur, this was a major step in the right direction compared with previous groups of primitive sprawling reptiles. For some time early in the twentieth century, several palaeontologists held the view that birds must have evolved from some of these 'thecodontian' reptiles, or early archosaurians, because certain features in their skeleton indicated they must have split off from the reptile family tree well before the first dinosaurs appeared. The fossil evidence for this theory was slim and it would take many new dinosaur discoveries in the second half of the twentieth century to take the theory of bird origins in a new direction.

In 1969 American palaeontologist John Ostrom at Yale University completed his study of *Deinonychus antirrhopus,* an extraordinary little predatory dinosaur only two metres in length. The carnivorous dinosaurs are known as theropods and *Deinonychus* is an excellent example of this group. Ostrom recognised in its agile form and deadly retractable claws that it must have been a fast-moving predator, so supported the then radical idea that dinosaurs could have been warm-blooded creatures. In *Deinonychus* he also saw many similarities with birds and became a fervent supporter of Huxley's original theory that birds had evolved from theropod dinosaurs.

From about the 1980s onwards, the cladistic revolution was a major change in the way scientists analysed fossils to determine evolutionary relationships between groups of animals. Hypotheses were structured around the types of characters exhibited by the fossil, such as the shape and special features of certain bones. Instead of trying to find suitable ancestors, or hypothetical links to join one group, say dinosaurs, with another group like birds, the detailed character states seen in the anatomy of these groups were rigorously analysed using new methods. The way of doing this was to feed a matrix of the characters and species showing the character states into a computer program, which parsimoniously analysed the data and produced a tree. The tree depicts the evolutionary relationships between 'clades' or groups of species, and these trees are therefore called 'cladograms'. Each cladogram changed with every new discovery as new fossil discovery added new character states to be incorporated into the current hypothesis. Sometimes the new find would slot in and not change the tree very much, but at other times radical discoveries could shake the tree vigorously, letting it settle again as a new configuration just a little more solid than before.

In the mid-1980s Jacques Gauthier of the University of California, Berkeley, identified features unique to birds and certain groups of predatory dinosaurs and fed these into his analyses. This work began to solidify the bird-dinosaur hypothesis, with other less robust hypotheses fading into obscurity. Today we see that the evolution of birds from dinosaurs is one of the most accepted and unchallenged hypotheses of vertebrate evolution, with strong support evidenced by a large suite of special anatomical characters found only in birds and some theropod dinosaurs. These features define not only the pattern of skeletal evolution between birds and dinosaurs, but also frame a set of linked physiological features (testable in modern birds and reptiles) that reveal the transformational stages from a running dinosaur to a flying bird.

Recent discoveries of complete, articulated fossil birds from the age of dinosaurs in the past 10 years or so have greatly altered our perspective on early bird diversity. From less than a dozen known fossil birds of this age in the 1960s, we now have at least 70 species, including a wide variety

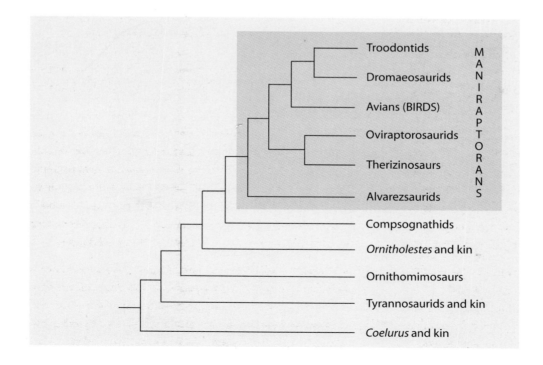

A cladogram showing the evolutionary relationship between the various coelurosaurians.

of forms that clearly occupied a range of ecological niches. While the dinosaurs and pterosaurs were the largest of all creatures on land and in the sky respectively, the smaller mammals and birds were evolving another more desired attribute – a degree of ecological flexibility which gave them an inherent ability to adapt and survive in the shadow of their dominant reptilian masters. Such traits served them well when the final apocalyptic event occurred that caused the great mass extinction 65 million years ago. At this time a gigantic meteorite or asteroid estimated at 10 kilometres in length came crashing down at 11 kilometres a second into the Gulf of Mexico at Chicxulub. The ensuing environmental chaos saw the ultimate demise of dinosaurs, pterosaurs and some of the great marine reptiles, whilst giving passage to the survival of birds, mammals and a number of a smaller reptilian and amphibian groups.

Today, birds are the most diverse of all terrestrial vertebrates with approximately 10 000 species known. We know of close to 5000 mammal species, whereas our greatest estimation of dinosaurs, based on the fossil record, is around 1000 species, although there is no doubt in any palaeontologist's mind that many more species would have existed. Currently around 20 to 30 new dinosaurs are discovered and named each year.

In order to comprehend the origins of birds we first need to learn about the major groups of dinosaurs and how they are classified into their family groups. Dinosaurs are classified into two main orders, the saurischians (meaning 'lizard-hipped' forms) and ornithischians ('bird-hipped' forms). The ornithischians include many of the armoured and horned groups of plant-eaters, such as the stegosaurs and ceratopsians, the ankylosaurs and the head-butting pachycephalosaurs, as well as the 'cows' of the Cretaceous, the hadrosaurs and hypsilophodontids. Although called 'bird-hipped dinosaurs' the ornithischians have no direct relationship to birds at all. It was the lizard-hipped dinosaurs, or saurischians, from which birds evolved.

Saurischians include two major groups, sauropods and theropods. The mighty, long-necked sauropods such as *Brachiosaurus* were the largest creatures to ever walk the Earth. The theropods were the carnivorous dinosaurs and include a variety of forms ranging from the mighty *Tyrannosaurus* down to the pigeon-sized *Microraptor*. Theropod means 'beast-footed', as the early anatomists saw resemblances to the feet of mammals. The story of the evolution of birds is tightly nested, so to speak, within this group of largely carnivorous dinosaurs. From the oldest and most primitive theropods such as *Herrerasaurus* from the Middle Triassic of Argentina, 228 million years ago, we start to see the appearance of bird-like features in the dinosaur skeleton. One of the primitive groups of theropods is the coelurosaurians, meaning 'hollow lizards', referring to the hollow nature of their vertebrae, a condition that would carry on to all birds. It is with this group that we shall start our narrative of this great evolutionary event.

2

MIGHTY TYRANNOSAURS AND THEIR HUMBLE ANCESTRY

We're lucky to have the opportunity to know T. rex, *study it, imagine it, and let it scare us. Most of all, we're lucky* T. rex *is dead.*[4]

The coelurosaurians include a wide range of predatory dinosaurs from the two-metre *Coelurus* to the gargantuan 14-metre-long killer *Tyrannosaurus*. The evolutionary divergence of tyrannosaurs from the small coelurosaurs is now much better understood since the recent discoveries of small, early tyrannosaurs such as *Eotyrannus*, and the Chinese feathered form *Dilong*. The most primitive members of the group, such as *Coelurus fragilis*, come from the late Jurassic (around 150 million years ago), so are contemporaries of the first bird, *Archaeopteryx*.

Coelurus has been the subject of much debate as to where it fits in the evolution of the tyrannosaurid dinosaurs, mainly because it is only known from a partial skeleton missing most of the skull. Nonetheless dinosaur experts tend to agree that it fits somewhere near the beginning of the feathered dinosaur story. It is a generalised theropod that shows the humble beginnings of the adaptations required for the evolution of birds. First, it has light, hollow bones and, second, it and all its kin (as contained in the rest of this book) have relatively larger brains than other more primitive meat-eaters such as *Allosaurus* and the strange Gondwana group known as abelisaurids.

The line leading to *Tyrannosaurus* might have started with moderate-sized forms like *Tanycolagreus*, a three- to four-metre-long, lightly built hunter with long grasping arms and fingers. It shares certain similarities in its pelvic bones and femurs with early tyrannosaurids.

The first undoubted tyrannosaur is the three-metre-long *Guanlong* from the Jurassic of China. It sported a strange crest on its head but is intimately linked to *Tyrannosaurus* and its kin by having

---

4 John Horner & Don Lessem (1994), *The Complete* T. Rex.

a peculiar U-shaped cross-section of its teeth in the upper jaw, and the bones of the snout (nasals) fused into a single bony unit. It was clearly a secondary predator in its environments, side-stepping the likes of the 10-metre-long *Sinraptor*.

Dilong, another early tyrannosaurid from China, shows undoubted presence of feathers adorning the tail. Although small (about 1.6 metres in length) it is more advanced than *Guanlong* in having a skull more like the later tyrannosaurs in many specialised features, but particularly by the hollow cavities in many of the bones (termed 'pneumatisation').

During the Early Cretaceous a range of small primitive tyrannosaurs (three to five metres in length) including *Eotyrannus* stalked the jungles of South-East Asia and Europe. *Siamotyrannus* from Thailand was once thought to be an early tyrannosaur but this has since been discounted due to lack of diagnostic skeletal remains for it to be confidently placed within this famous family.

The main radiation of tyrannosaurs is typified by the many strange forms that appeared in the Late Cretaceous of North America and Asia. Most were large (eight to nine metres) to very large (up to 14 metres) with puny forelimbs having only two fingers on each hand. Most bear some form of rugged ornamentation on the top of the skull, ranging from a series of sharp cones along the front of the snout of *Alioramus*, through to roughened, knobbly surfaces above the eyes, possibly for attracting mates or for head-pushing contests between rival males.

The primitive tyrannosaur *Eotyrannus* stalked the jungles of Europe.

The first relatively complete skeleton of *Tyrannosaurus* was discovered in 1902 in eastern Montana, USA, and was described by American palaeontologist Henry Fairfield Osborn in 1905. After several rigorous field seasons, Barnum Brown, Osborn's field collector, had brought back a number of partial skeletons (eight specimens had been found by 1912), and by 1915 the American Museum of Natural History was able to mount the world's first complete skeleton of *Tyrannosaurus* in its halls. The late, great Stephen Jay Gould publicly attributed his desire for studying palaeontology down to the time his father took him to see the *T. rex* in the American Museum of Natural History when he was five years old.

Tyrannosaurus and its close Asian cousin *Tarbosaurus* reached lengths of 12 to 14 metres, with estimated body masses of around six tonnes. Most tyrannosaurs however were more moderately large animals: *Daspletosaurus*, *Gorgosaurus* and *Albertosaurus* reached eight to nine metres in length and weighed between one and two tonnes. The group were restricted to the Eurasian landmass, comprising North America and Asia.

Although popular movies such as *Jurassic Park* will have us believe *Tyrannosaurus* was a swift runner ('we've clocked the *T. rex* at 32 miles per hour') reliable studies of the physiological constraints of its large weight and degrees of limb movement suggest more modest speeds of around 10 miles per hour were close to its maximum. Such studies lead some palaeontologists such as Dr Jack Horner to suggest that *Tyrannosaurus* and its kin were either largely scavengers or ambush predators that surprised their prey. Others believe that the prey of *Tyrannosaurus* was probably similarly slow-moving, so it was likely to have been an active hunter.

Artist's reconstruction of the skeleton of *T. rex*.

New studies on *Tyrannosaurus* show that it grew quickly, although its rate of growth accelerated in its later years. A 10-year-old *Tyrannosaurus* may have weighed only half a tonne. Its major growth spurt came on between 13 years (weight at one tonne) through to 20 years of age (reaching five tonnes) and it may have lived to be 30 or so years of age, peaking at around five-and-a-half tonnes. Although fossilised soft tissues have been extracted from its bones, no DNA material has been successfully located from the fossils, and at 65–70 million years old, it is highly unlikely that *Tyrannosaurus* DNA will ever be found preserved intact enough to study even small segments of its genome.

Tyrannosaurus and its kin have fired the imagination of people of all ages who have seen the real skeletons in their dignified museum poses. The new vision of *Tyrannosaurus* is to see it as a close ancestor to modern birds. Coelurosaurian dinosaurs such as *Tyrannosaurus* had a V-shaped wishbone, three main toes facing forwards, and long arms that had lost the fifth and fourth fingers on the hand. To tell the rest of this story we must move on to the more advanced theropods, the ones which begin to resemble bird-like creatures.

3

FLEET-FOOTED OSTRICH MIMICS

Indeed the comparisons of this dinosaur in various aspects of its anatomy to an ostrich are so striking that it is often called the 'ostrich-like dinosaur' which, of course, does not imply any direct relationships but does indicate that Ornithomimus, *during the Late Cretaceous times, lived a life that is paralleled by the large flightless birds of today.*[5]

The ornithomimosaurs are best thought of as the stream-lined sporty models built mainly for speed. They were a group of theropods whose life is thought to have been much like the flightless ratite birds today (such as the ostriches, emus and rheas), being mostly omnivorous, although one had stomach contents preserved, showing it was largely herbivorous.

In recent years the discovery of the earliest and most primitive members of this group have shed much light on the origins and evolutionary radiations of the group. They are now considered as fairly primitive within the hierarchy of advanced theropods, resting comfortably between the primitive coelurosaurians and the more advanced group called the 'maniraptorans' such as the therizinosaurs, alvarezsaurs, oviraptorosaurs, troodontids, dromaeosaurs and birds.

The first ornithomimosaur fossils were found in 1889 in the late Cretaceous rocks near Denver, Colorado, by George Cannon. Charles Othniel Marsh named these fragmentary foot bones as *Ornithomimus* meaning 'bird mimic' as they looked so much like that of a bird's foot. It wasn't until 1917 that Lawrence Lambe, hunting along the Red Deer River outcrops of Alberta, uncovered the first relatively complete skeleton which Henry Fairfield Osborn later named *Struthiomimus* meaning 'ostrich mimic'. Osborn perceptively noted his dinosaur had many similarities to the fast-running ratite birds of today.

---

5   Edwin Colbert (1969), *Evolution of the Vertebrates*, 2nd ed. p. 201.

The most significant series of discoveries of new ornithomimosaurs came about in the 1960s Polish–Mongolian dinosaur expeditions, and in the 1990s from newly discovered sites in Spain and China. The long, stiff, horizontal tail, elongated legs for fast running, and lightly built, small skull perched atop long, gracile necks indicates they were built to run fast. Fossilised trackways attributed to ornithomimosaurs suggest maximum speeds of up to 36 miles per hour, making them the swiftest of all known dinosaurs, and equally as fast as our modern racehorses.

The primitive ornithomimosaur, *Pelecanimimus*, had numerous small teeth.

The most primitive member of this group is *Pelecanimimus* from the famous Early Cretaceous lake deposits at Las Hoyas in Spain. *Pelecanimimus* (meaning 'pelican mimic') is the only member of this group to have numerous small teeth, some 220 in all. Some other ornithomimosaurs retain a few teeth at the front of the jaws only, such as *Shenzhousaurus* from Shenzhou in the Liaoning Province of north-eastern China, and *Harpymimus* from Dundgov, Mongolia. The remaining members of the ornithomimosaur clan lack teeth and have well developed beaks. Some forms have been found with small rounded stones called 'gastroliths' in their stomachs. These were probably used to help digest their food. In modern creatures only herbivores have such stones to help grind up hard vegetative material in the gut, such as seeds, which accords well with the finding of vegetation in the stomach region of at least one member of the group, *Sinornithomimus*.

One theory had it that some ornithomimosaurs may have fed like flamingos, filter feeding the water with their curved beaks, but other specialists have disputed the evidence for this, instead saying that structures in their palate were for supporting the beak only, and not for specialised filter feeding. We do know that some ornithomimosaurs protected their young by herding or gregarious behaviour. A collection of some 14 fossilised skeletons of *Sinornithomimus* found together in the Inner Mongolian autonomous region of China, showed that 11 of the 14 were juvenile or sub-adult animals, so the three adults with them were most likely protecting the group of young ones.

The most specialised of the ornithomimosaurs are divided into a North American group comprising *Struthiomimus*, *Dromiceiomimus* and *Ornithomimus,* and an Asian group of *Gallimimus* and *Anserimimus*. *Ornithomimus* shows the typical form of the advanced ornithomimosaur skull in having large eyes, a slender, curved snout with matching lower jaw that bends downwards at the front. The large orbits of ornithomimosaur skulls are often preserved with big sclerotic bones, which surrounded the eye in life, and indicate a highly developed sense of vision. The large brain size of ornithomimosaurs is comparable with those of modern birds and the very brainy troodontid dinosaurs. Although the group is typified by having long legs and long gracile arms, the mechanics of the forelimbs in *Struthiomimus* suggest that they were not equipped for digging or raking, but best suited to grasping branches and fern fronds.

4

ORNITHOLESTES, 'COMPYS' AND KIN

It is impossible to look at the conformation of this strange reptile and to doubt that it hopped or walked, in an erect or semi-erect position, after the manner of a bird...[6]

This group covers a variety of mostly small, enigmatic dinosaurs such as *Ornitholestes* and other small dinosaurs thought to be closely related to it, as well as the family Compsognathidae. The compsognathids were all relatively small, gracile dinosaurs that most probably hunted insects, small lizards, frogs and other like-sized prey. They include small forms such as *Compsognathus* (the 'compys' of *Jurassic Park* fame), *Juravenator* and *Sinosauropteryx*, the latter being the first feathered dinosaur discovered and announced to the world in 1996.

Ornitholestes was discovered at Como Bluff, Wyoming, in 1900 and described in 1903 by Henry Fairfield Osborn of the American Museum of Natural History. For many years it was depicted with a strange horn on its nose, similar to *Ceratosaurus* from the same deposits. Osborn imagined this agile little dinosaur sneaking up and catching birds like *Archaeopteryx* – its name enigmatically means 'bird robber'. A recent paper by Kenneth Carpenter and colleagues that redescribed the original material of *Ornitholestes* has shown that the skull did not really have a short horn on its snout, instead the skull bones had been displaced due to compaction in this region.

Proceratosaurus, known from a partial skull discovered in 1910 in Gloucestershire, England, did have an elaborate crest along its snout – possibly for signalling to mates. *Nqwebasaurus* (meaning 'lizard from Nqweba') is a small, almost complete dinosaur found in South Africa and described in 2004 by Billy De Klerk and others. Although fairly nondescript (as were most of these 'coelurosaurians'), *Nqwebasaurus* apparently had partially opposable thumbs so had some kind of special manipulative features in its hands for grasping prey.

The compsognathid family takes its name from *Compsognathus* (meaning 'delicate jaw'), discovered in 1850 by Dr Joseph Obendorfer from the Solnhofen lithographic limestones of

---

6   Thomas Henry Huxley (1867), p. 361.

Bavaria, Germany. For many years it was thought to be the smallest dinosaur – it was only around a metre in length – although we know now of many other theropods around this size or smaller. It was formally described by the German Andreas Wagner in 1861 and studied by British anatomist Thomas Henry Huxley in 1867. He was particularly enamoured with *Compsognathus*, regarding it as 'the most bird-like dinosaur' known at the time. The remains of its last meal, a lizard (*Bavarisaurus*) was identified inside the stomach of the original *Compsognathus* specimen by Professor John Ostrom, showing it must have been a reasonably swift moving predator. One interpretation of it as a possible fish-eating, swimming dinosaur by the late Bev Halstead in his 1975 book *The Evolution and Ecology of Dinosaurs* was a fanciful idea, but has no hard support from its skeletal anatomy. A second complete specimen of *Compsognathus* is known from France and shows it grew to around 1.4 metres.

Juravenator was another small compsognathid discovered from a limestone quarry near Schamhaupten in Germany and nicknamed 'Borsti' by its finders, alluding to the possibility it should have had bristly feathers covering its body. Unfortunately the well-preserved specimen did not show any feathers attached to it.

Huaxignathus is known from an almost complete specimen found in northern China and it was formally described by Sunny Hwang and others in 2002. It is regarded as the most primitive of all compsognathids as it lacks the specialised features of the forearms seen in other species.

Sinocalliopteryx is the largest known dinosaur fossil with feathers attached to it.

Arguably the most significant dinosaur find of the past century was the 1996 announcement of the little compsognathid *Sinosauropteryx prima*, from the remarkable Liaoning deposits of north-eastern China, as it had what appeared to resemble a furry covering to its entire body that resembled simple feathers. The first photos of it were shown clandestinely at a Society of Vertebrate Paleontology meeting in the USA, and this prompted a visit to China by a group of leading US and German experts, who spent three days poring over the specimen to determine if indeed it really had feathers. Although the group did not at the time conclude it definitely had feathers, further detailed study of similar well-preserved finds from the same site have confirmed that *Sinosauropteryx* was indeed covered with hair-like proto-feathers.

The most amazing thing about the *Sinosauropteryx* fossil was the discovery inside of the jaw of an early mammal that had made up its last meal! Such extraordinary fossils that reveal a wealth of information about dinosaur biology do not come along very often.

Most recently a new discovery from China has made us change the way we have long thought about compsognathids as agile lizard-hunting dinosaurs. The 2.4-metre-long skeleton of *Sinocalliopteryx* revealed that its last meal was the leg of a dromaeosaurid. However the most interesting thing about this giant predatory 'compy' is that it had feathers covering the body and legs, with slightly enlarged feathers around the ankles, similar to, but not as well developed, as in the 'four winged dinosaurs', the *Microraptor* group. It proved that the existence of feathers on the legs was a trait that went right back in dinosaur evolution to the primitive coelurosaurs. It demonstrates admirably how the discovery of one new spectacular dinosaur fossil can provide new insights on the evolutionary transition between dinosaurs and birds.

5

THE MONSTER CLAWS

Each digit on the three-fingered hand ends in a fearsome claw that can reach almost one meter (3 feet) in length, the largest claws of any known animal.[7]

Perhaps the most unusual-looking beasts of the dinosaur world were the therizinosaurs – the name means 'reaping lizard'. They may have been the equivalent of reptilian giant sloths of the late Mesozoic Era and remind us of the ghoulish 'Freddy' from the *Nightmare on Elm Street* movies in having hands armed with razor-sharp claws up to 70 centimetres long. For many years little was known about these enigmatic dinosaurs apart from scant remains, but in the past decade the first relatively complete skeletons of therizinosaurs have been discovered, confirming that these dinosaurs were every bit as bizarre as they were first imagined.

The discovery of this strange group was made in the 1940s by a joint Soviet–Mongolian dinosaur expedition, which retrieved three huge bony claws, some up to 70 centimetres in length, thought to originally belong to a giant turtle and so named *Therizinosaurus cheloniformis*. *Erlikosaurus*, also from Mongolia, was the next therizinosaur to be described, by palaeontologist Perle Altangerel in 1981, based on a skull and some bits of the skeleton. However the first relatively complete skeleton of a therizinosaur was *Alxasaurus* described from Mongolia by Canadian dinosaur specialist Dale Russell and legendary Chinese dinosaur expert Dong Zhiming in 1994. They showed that *Alxasaurus* bore a small head with many crudely serrated leaf-shaped teeth, a robust, deep body characterised by large pelvic bones, and long arms equipped with unusually long claws. The discovery of other partially complete therizinosaur skeletons soon followed with *Neimongosaurus*, *Nothronychus* and *Falcarius*. The overall picture of the group emerged as them all having a relatively slender, elongated neck, a small head with many leaf-shaped teeth, rotund thick bodies, and large hand claws, indicating they were not directly carnivorous like the tyrannosaurids, but probably more omnivorous or herbivorous.

Like the ornithomimosaurs, the therizinosaurs seem to have been a group of initially predatory theropods that secondarily shifted to herbivory. The teeth as such have close resemblances to the

---

7   Henry Gee & Luis Rey (2003), *A Field Guide to Dinosaurs*, p.135.

plant-eating dinosaurs such as hypsilophodontids or some ankylosaurs, and the enlarged shape of the body is seen as a typical adaptation in many animals for holding large amounts of plant material in the gut.

The most primitive member of the group is *Falcarius* from the Early Cretaceous of Utah. It had many leaf-shaped serrated teeth, closely packed together on its jaws, as a well as a host of skeletal features, particularly in the pelvis, that are modified in all later members of the group.

Beipiaosaurus had a toothless beak, suitable for cropping vegetation.

The most advanced forms of therizinosaurs have teeth more specialised for plant-eating as well as a toothless beak at the front of the mouth for cropping vegetation. The best preserved specimens come from the Liaoning sites in north-eastern China. *Beipiaosaurus* shows that, like many of the advanced theropods, therizinosaurs also had a feathery covering over the body. The feathers of *Beipiaosaurus* are found associated with the forelimb and near the pelvis. They consist of hair-like filaments up to 70 millimetres long, some having a hollow core.

Dinosaur eggs attributed to therizinosaurs are commonly encountered from the Cretaceous deposits of Henan Province, China. One of these prepared by Terry Manning of England appears to be an embryo of a therizinosaur. These show that the slender, elongated teeth in the jaws of the embryonic dinosaurs are later replaced by a more symmetrical type of tooth. The eggs of therizinosaurs are generally spherical in shape, although some gigantic, elongated eggs (up to 50 centimetres in length) have also been identified as belonging to large therizinosaurs by Canadian palaeontologist Phil Currie.

Their robust, heavily built skeletons, small head and giant claws suggest that these dinosaurs had a sloth-like lifestyle. Perhaps they used their giant scythe-like claws to grip and shred leaves off trees, or for scrapping the trunks of certain trees to eat the outer bark covering and get to the sap. The nature of their teeth and jaws shows they were comfortably vegetarian. Much like the panda bears of today, they are examples of once primarily predatory animals that adapt to eat plants when an abundance of certain food types needs to be exploited. The increasing rise of flowering plants, angiosperms, in the later part of the Cretaceous Period, might well have been the main trigger for the radical shift in therizinosaur diet.

6

THE EGG BROODERS

Remains of a 128 million-year-old buck-toothed dinosaur that looks like a cross between a rabbit and a prehistoric creature have been found in China, scientists said on Wednesday[8]

Oviraptorosaurs would have to be amongst the weirdest looking of any creature to have ever evolved. Some of them, like the decidedly odd *Incisivosaurus*, as the *China Peoples Daily* newspaper claims, might have had a similar lifestyle to our modern rabbits, using their enlarged front buck-teeth teeth to gnaw down vegetation. Despite their exotic appearance, the fascinating assemblage of oviraptorosaurs provides rare insights into the habits or brooding and nesting dinosaurs from remarkably well-preserved fossils coming out of the Mongolian deserts. Yet our scientific study of this peculiar group of animals started only about 80 years ago.

In the early 1920s an expedition lead by Roy Chapman Andrews of the American Museum of Natural History explored much of central China and Mongolia in the search for mankind's origins. Despite not uncovering the kinds of human fossils they hoped to find, they stumbled upon remarkable finds of dinosaur skeletons and the very first nests of fossilised dinosaurs eggs.

One specimen found in 1923 by George Olsen showed a fossilised dinosaur skeleton found near a nest of eggs, so the finders immediately thought the worst – that this dinosaur had died in the very act of trying to steal another dinosaur's eggs. The eggs were thought to belong to a ceratopsian or horned dinosaur. This interpretation, reinforced by its strange skull with almost toothless jaws, spawned the ill-fitting name *Oviraptor philoceratops* for the dinosaur, meaning 'egg stealer, fond of ceratopsian eggs'.

Discoveries of complete oviraptorid skeletons in the late 1980s and early 1990s finally vindicated the animal by showing they were actually sitting atop their own nests, brooding their clutch of well-arranged eggs like modern birds.

---

8 *China Peoples Daily*, 19 September, 2002.

Oviraptorosaurs remain amongst the strangest looking of all theropods. The skeleton of *Oviraptor* was relatively complete and showed an animal up to two metres in length with large forearms and sharp claws, but the skull was short and built with many openings and a strange crest on top of the head. The most remarkable discovery from the early studies of the group was the identification of a bone that turned out to be a furcula, or wishbone, as occurs in the breast skeleton of birds. This was a feature then not known in other theropods. The sides of the furcula were fused together as a single V-shaped bone at the front of the chest.

The discovery of well-preserved, new oviraptorosaurs from Mongolia and Liaoning, China, in recent years has added much new information about this extraordinary group. They are most commonly found in the Early to Late Cretaceous deposits of Asia with a few forms known from the Late Cretaceous of North America. The most primitive member is *Incisivosaurus*, the buck-toothed dinosaur whose enlarged, flat, front teeth look much like the incisors of gnawing mammals. *Incisivosaurus* had a few other small teeth along its jaws whereas all other known oviraptorosaurs lack teeth entirely, except for *Caudipteryx* and *Avimimus,* which retain a few teeth on the front of the upper jaw. Although most oviraptorosaurs were small animals under two metres in length, the recently discovered *Gigantoraptor* from China shows that some members of the group could grow up to eight metres in length.

Caudipteryx (meaning 'tail feather') is now known from nearly a dozen relatively complete individuals found since the late 1990s from the famous Liaoning lake deposit of north-eastern China. It had an exceptional covering of feathers surrounding its body with larger plumes emanating from its wings and fanning out from the rear half of its tail. The short arms and long legs of *Caudipteryx* are indicative of a fast-running animal, and the presence of feathers are suggested as being primarily for display.

The pygostyle, a fused series of vertebrae that forms the standard tail bone of modern birds (supporting the 'parson's nose' of your roast chicken), was identified in an oviraptorosaur from Mongolia by Rinchen Barsbold and colleagues in 2000, indicating an even closer relationship of this group to birds than previously thought.

The skulls of oviraptorosaurs often show strange bony crests adorning the top of the head, best exemplified by forms such as *Rinchenia* and *Nemengtomaia* from Mongolia. These crests are formed by the enlarged and often fenestrated nasal bones. The nasal openings are thus seen to be placed in somewhat bizarre positions, such as above the eyes and behind the beak-like structures. The nasal bones of most oviraptorosaurs are fused together. In some respects these look similar to the crest of modern cassowary birds, but as the Mongolian oviraptorosaurs lived in sandy but humid environments, one cannot imagine them using the crests to part the thick vegetation in the same way as the jungle-dwelling cassowary. More likely the crest was used mainly for courtship displays, enhanced by the spectacular displays of feathers adorning the arms and tail.

We have little evidence of what oviraptorosaurs actually ate apart from the remains of a very small theropod skeleton found associated with one oviraptorosaur nest, implying it was brought back as food for the young ones. Together with their horny, toothless beaks and sharp hand claws, it

Incisivosaurus was a primitive oviraptorosaur with strange buck-teeth.

is likely that the group foraged for a wide range of prey items, including small animals, eggs and some plant foods. The presence of gastroliths in the gut of *Caudipteryx* implies it may have been largely herbivorous.

The extraordinary preservation of oviraptorosaurs sitting atop their nests of eggs is seen in some of the Mongolian sites where large wet sand dunes may have occasionally collapsed upon brooding dinosaurs like *Citipati*, killing them instantly. The nests show that the dinosaurs laid their eggs two at a time in a circle, before laying another smaller circle of eggs on top of them. Up to three layers containing some 30 eggs occur in some nests and the brooding dinosaur has been found with arms outstretched over the nest, possibly shading them to regulate the amount of heat in the sand.

7

THE DEADLY RAPTORS

They show extreme intelligence, even problem solving. Especially the big one. ...
That one ... when she looks at you, you can see she's thinking (or) working things out.[9]

No dinosaur has better captured the gothic side of human imagination than the cunning raptors depicted in the *Jurassic Park* movies. Rather than presenting us with slow, dull-witted lizards as in the early Hollywood classics such as Cooper and Schoedsack's *King Kong*, director Steven Spielberg's raptors depicted dinosaurs in a new and disturbing image. With the power of hundreds of scientific experts restoring them, and embellished with the imaginations of Hollywood scriptwriters, the *Jurassic Park* raptors come across as fast, agile and highly intelligent predators.

In many respects this is still the way that scientists think of dromaeosaurids, with good reason to do so, as the fossil evidence backs up much of the inferred behaviour of the group presented in the movies. The voracious pack-hunting behaviour shown by the *Velociraptor* in *Jurassic Park* was actually based on the discovery of several two-metre-long *Deinonychus* specimens found in Montana around the skeleton of an eight-metre prey animal, a *Tenontosaurus*. Although the geological context of the site could alternatively mean that they were all washed together as carcasses by the same stream action, the accepted interpretation (especially in popular science books) is to see this association as direct evidence that the raptors were pack hunters that worked together, like modern wolves, to bring down the bigger plant-eating dinosaur. No other group of raptors has been found associated in the same geological deposit.

The first discovery of a member of this infamous 'raptor family' (Dromaeosauridae) was from the Red Deer River district of Alberta, Canada, in 1914 by ace dino hunter Barnum Brown. It was later named *Dromaeosaurus,* meaning 'running lizard', because of its light, agile build. The more complete remains of another species were found not long afterwards during the American-Asiatic expeditions in the 1920s. This more complete skeleton showed that it had peculiar retractable sickle-shaped claws on its hands and on the second toe of the foot. The creature was about two metres long. It was named *Velociraptor* (meaning 'speedy thief') by Henry Fairfield Osborn in

---

9   Muldoon, game keeper, from the movie *Jurassic Park* (1993), directed by S. Spielberg.

1924. The Polish–Mongolian expeditions of the 1960s later uncovered one of the most famous and spectacular of all dinosaur fossils, the battle between a *Velociraptor* and a *Protoceratops*, frozen in time after a large amount of unstable wet sand collapsed upon them.

We now know of around a dozen or so members of this clan, with exceptionally preserved examples coming out of northern China. Their bodies were covered by a down of short filamentous feathers whilst some had expanded plumes of pennaceous feathers on the arms and some, like the half-metre-long *Microraptor*, even had them on the legs, giving a four-winged appearance to the animal.

The dromaeosaurids are principally defined by the special anatomical features of their skull bones and jaw joint. However, their deadly retractable foot claws, which are much better developed as weapons than in the troodontids, are accepted as their hallmark feature. Dromaeosaurids also have long, rod-like ossifications attached to their tail vertebrae, making the tail rigid and inflexible.

Microraptor had feathers on both arms and legs, giving it a four-winged appearance.

Most dromaeosaurids were small animals, between half-a-metre and three metres in length. Some reports of giant raptors, such as the six-metre *Megaraptor* from South America, remain inconclusive in the eyes of leading dinosaur experts, so the largest member of the group we know of for sure is *Utahraptor* from the Early Cretaceous of Utah, growing up to seven metres in length and weighing around 700 kilograms.

Dromaeosaurids had a unique deadly weapon on their hands and feet – razor-sharp retractable claws. At first palaeontologists thought these weapons might be primarily used for disembowelling prey, but tests carried out on the strength and cutting ability of the claws proved otherwise. Their claws are now regarded as a climbing tool to help the raptors rapidly scale the backs of larger prey dinosaurs so they could viciously attack the neck and spine. The use of such hooked claws, combined with winged arms, might well have been a combination that could have lead some of them to climbing trees. However, as yet we have no direct evidence suggesting they did as, for example, we would find in the specialised hands or feet of modern arboreal animals.

The sleek, sporty build of the dromaeosaur, with its long legs and counterbalancing tail shows clearly it was an animal with a highly active metabolism which, along with a large brain size, well-developed branching feathers and very long arms, are all features shared with birds. The dromaeosaurids could also fold their arms back in a way similar to how modern birds fold their wings back. Some palaeontologists argue that dromaeosaurids are the closest relative to birds and that birds arose from within this group. Others hold that troodontids might be more closely related to birds. However, a recent theory is that dromaeosaurids and troodontids are together the group that is most closely related to birds, and this idea explains why both of these small dinosaur groups share some features in common with birds.

8

THE LITTLE SMART ONES

This specimen [of the troodontid Mei long*] displays the earliest recorded occurrence of the stereotypical sleeping or resting behaviour found in living birds.*[10]

A breath-taking fossil was unveiled to the world on 13 October 2004. Named *Mei long,* meaning 'soundly sleeping dragon', the little dinosaur measured only half-a-metre in length and died from being covered by volcanic ash whilst it was asleep. The most extraordinary thing about the fossil skeleton was that the dinosaur slept in a curled up posture, today seen in living birds. Here the fossil remains provided more than just anatomical evidence linking dinosaurs to birds, but also strong behavioural evidence. *Mei long* was one of the troodontid dinosaurs, a group some scientists argue are closer to birds than any other theropod lineage. The troodontid group, named after the first discovered member of the family called *Troodon,* has been known for over 150 years, but from rare scant remains. Today, with such remarkable new finds coming from China, they are much better understood and their inferred lifestyles clearly show strong parallels with birds.

Troodontids were first known from teeth found in Montana, described and named as a lizard, *Troodon,* by Joseph Leidy of North America in 1856. However the group did not emerge as a distinct family of theropods until more complete remains of *Troodon* were collected and shown to belong to a theropod dinosaur. The detailed description of its skeleton in the 1980s by Canadian palaeontologist Phil Currie and colleagues elucidated its anatomy and demonstrated a close alliance of the genus with birds. However, the majority of well-preserved troodontids have come out of Mongolia and northern China, where new evidence has emerged showing not only that they were feathered but also how some of them, like the little *Mei long,* slept in a resting position similar to that of a modern bird.

Troodontids were small dinosaurs, most being a metre or less, and none being larger than two metres or so. They had brains more developed and larger relative to their body size than any other dinosaurs, with particularly enlarged areas for sight and hearing. Some had elaborate

---

10 X. Xu and M. Norell (2004), *Nature* 431, p. 840.

arms covered in bird-like plumages. Indeed some of them are so close to being birds that some of their fossils were first described as being birds, such as the feathered *Jinfengopteryx* from northern China, now regarded as most likely being a primitive troodontid. Studies of the bone sections of *Troodon* indicate that these animals grew quickly, reaching maturity in just less than five years.

One of the most primitive members of the troodontid family is *Sinovenator* from Liaoning in north-eastern China. It shows that some of the key structural modifications in the evolution towards birds were already developed in the primitive troodontids. The endocast of the braincase is strikingly similar to that of the first bird, *Archaeopteryx*, as is the presence of several hollows in the base of the braincase (known in technical jargon as the 'periotic pneumatic systems'). The distinctive teeth of *Troodon*, with coarse serrations along each edge, proved to be anomalous, as other troodontids have the leading edge of the tooth as a sharp, knife-like blade, with the rear edge of the tooth being coarsely serrated. The teeth of *Byronosaurus* lack serrations and are more like those of early fossil birds. Like the more infamous raptors dromaeosaurids, troodontids also share a retractable sickle-shaped claw on the second toe of the foot.

Mei long is one of the smallest troodontids, measuring only 53 centimetres in length, yet represents a mature individual as the bones are well fused together in the skull. Its sleeping posture shows the body resting on folded legs aligned in a neat symmetrical pattern. The arms are also symmetrically arranged and splayed outwards relative to the body. The little head was tucked inside the folded arm, similar to the way modern birds sleep. It slept on the ground though, not in a nest in a tree.

The evolution of the troodontids is poorly known as there are fewer than a dozen known species, and the remains of only some of them are relatively complete. It seems that the earliest forms already had a lighter skull, and were equipped with a well-developed sense of sight, hearing and balance, perhaps used most effectively in their daily life chasing prey. Maybe such specialisations were their key innovation, either by using their enhanced sensory skills and sharp sense of balance to enable them to leap down upon prey from trees with feathered wings to aid them, or to successfully chase and catch agile prey items such as large insects or birds.

Some features of the skeleton of the primitive troodontid *Sinovenator* show clearly its relationship to birds.

9

BIRD DINOSAURS AND DINOSAUR BIRDS

As we discover more and more theropod lineages that acquired birdness, or evolved bird-like features, the line between what is a bird and what is not becomes increasingly blurred. [11]

In the early 1990s some isolated finds of strange bird-like dinosaurs from Argentina and Mongolia threw a spanner in the works of determining bird origins. The discovery of partial skeletons of *Alvarezsaurus* from Argentina and *Mononykus* from Mongolia revealed that the boundary between bipedal running dinosaurs and flightless birds was totally blurred. Then the discovery of more complete remains of a similar beast, *Shuvuuia* from Mongolia, revealed the true nature of these strange beasts. They had long legs, curved, delicate necks, small, gracile heads with tiny teeth, and short, powerful arms each with one very large claw and two remnant smaller claws.

At first the group was determined by Luis Chiappe and colleagues to be strange flightless birds. Almost immediately other scientists analysed the finds and argued that they were really dinosaurs. Further work and new discoveries saw them once again placed at the very beginnings of all birds. The final word from our good colleagues, including Luis Chiappe, is to now regard them as swift-running dinosaurs that are just outside the very hazy boundary demarcating the true birds.

The skeletons of the first birds looked so much like those of small predatory dinosaurs that even today it is difficult for dinosaur specialists to draw a clear line between them. In fact, it is entirely arbitrary where scientists choose to define the beginnings of birds. If, for example, the presence of feathers defined birds we would then have to include probably all the dinosaurs listed in this book as 'archaic birds'. Of course with skin and other integumentary features rarely preserved on most dinosaurs, it is well nigh impossible to tell if a fossil known from a few bones is really a feathered creature or not. Instead, it is far more practical to choose a combination of specialised

---

11 Luis Chiappe (2007), *Glorified Dinosaurs*, pp. 16–17.

features in the skeleton that are associated with the one ability with which we today largely define most birds – flight.

Most scientists working in this field now accept that *Archaeopteryx* is the appropriate creature to arbitrarily define the origins of birds, as it has wings capable of powered flight, feathers specialised for flight like a modern bird, and a few skeletal features found only in modern birds. It also has a suite of archaic features found in reptiles (such as teeth, separate wrist and finger bones, and several tail bones) that would be lost in later, more specialised lineages of birds.

Apart from *Archaeopteryx*, dated at around 150 million years old, few other birds were known from the age of dinosaurs for many years. Those that were came from latest Cretaceous sites in North America, and included strange sea-dwelling forms like *Hesperornis* and *Baptornis*. In the 1970s and 1980s a few isolated bones and occasional rare partial skeletons of dinosaur-aged birds shed a little more light on early avian evolution. Articulated specimens, such as *Iberomesornis* from the famous Las Hoyas site in Spain, demonstrated the presence of an early pygostyle in the tail, and a strut-like shoulder bone (coracoid), prerequisites for the fine-tuning of avian flight. The later discovery of *Eoalulavis* in the 1990s demonstrated how the wrist structure played an important role in birds acquiring specialised methods of flying with improved manoeuvrability.

However the true revolution in fossil bird discoveries, shedding significant light on their transition from archaic dinosaurian forms to modern birds really came about in the mid-1990s with the discovery of the now famous Liaoning lake deposits of north-eastern China. At first complete fossil birds from this site were smuggled illegally out of China and began to appear in auction houses in the United States and Europe, even before the fossils were studied and named. Eventually the significance of the sites was recognised and the better specimens were collected and studied by Chinese workers from various geological and palaeontologic institutes. The first to emerge from these studies was *Confuciusornis*, which today remains one of the best known Cretaceous birds, as it is represented by at least 1000 good specimens. The most remarkable thing about these fossils is their detailed preservation of complete skeletons with feathers attached. Some specimens show long tail feathers present whilst others lack these, suggesting sexual dimorphism was prevalent in early birds.

Confuciusornis had a beak lacking teeth and differed principally from *Archaeopteryx* in possessing a pygostyle, or shortened tail bone. *Changchengornis*, another close relative of *Confuciusornis*, had a smaller curved beak and more delicately constructed skull, and most importantly, a longer first toe that shows the beginnings of the grasping adaptation for perching.

In understanding the way in which modern birds fly we need to consider the economics of flight. For animals such as bats and even the extinct flying reptiles, the pterosaurs, flying was powered by the muscles of the arms combined with lightening of the bones. Pterosaurs most likely used winds and thermal currents wherever possible to glide for long distances. But birds do it far more easily, as all modern flying birds have a specialised system that enables the wing to flap with minimum effort. This was explained in late 2006 by David Baier of Brown University, USA, and

Confuciusornis was about the size of a modern crow.

his colleagues in the journal *Nature*. It is due to a highly specialised system utilising a powerful ligament, called the acrocorahumeral ligament (here shortened to the 'AHL'), that provides a mechanism for transmitting the force of the breast muscles through the coracoid bone in the shoulder which acts as a compressive strut. In other words, the bird pulls the wing down using its breast muscles, and the ligament system helps spring the wing back upwards without using additional muscular power.

Using well-preserved dinosaur and bird fossils it is possible to trace the origin of this specialised system back through time. We see that dinosaurs and crocodiles use a muscle-based system to balance the force used in the limbs. In advanced feathered dinosaurs such as *Sinornithosaurus* and *Sinornithoides,* the ligaments of the shoulder begin to move higher up the shoulder girdle, approaching the same position that eventually predominated in birds. The first birds such as *Confuciusornis* show the beginning of the AHL ligament starting to take over from other ligament and muscle systems used in the shoulder. Palaeontologists can see this clearly from the position of the facet where the arm bones meet the shoulder as the articulation has moved upwards relative to the coracoid bone. In modern flying birds the ligament runs parallel to the direction of the wishbone and thus represents an advanced stage of locomotory ability where the ligament is doing most of the work.

Birds have an efficient way of flying, thereby expending far less energy than their ancestral forms which put far more muscular power into their early flights.

The true success of modern birds lies not only in this system of ligaments assisting the flight stroke, but also in the accompanying sensory adaptations to process information faster during flight, and in their unique way of breathing, whereby the spent gases are diffused throughout the blood stream rather than exhaled through the mouth or nostrils like other animals. This system allows birds to breathe much faster as it's almost acting as a one-way system of air intake.

Today we have a multitude of beautiful birds, of many diverse shapes and sizes, numbering around 10 000 living species. Some lineages have secondarily shed their ability to fly for life on the ground. In doing this they trade off a high metabolic lifestyle for one that requires less fuel to survive. Some, like the penguins, use their primaeval flight stroke to propel them through the seas in search of fish.

Birds form a major part of the global human diet, and we love to keep them as pets, or simply enjoy hearing them sing in their wild habitats. But remember when you next tuck into a tasty morsel of chicken, duck or quail, that you are eating something that seamlessly evolved from the mighty predatory dinosaurs. There's just a hint of *T. rex* in every bite of chicken. Indeed, the recent findings of feathered dinosaurs and early birds show just how rapidly our perspectives on evolution can change. Birds today could be regarded as the supermodels of the dinosaur world. They are lasting icons of evolution, a true testimony to life's ability to change and adapt in the face of environmental chaos.

Changchengornis had a delicately constructed skull and toes adapted for perching.

THE DINOSAURS ILLUSTRATED

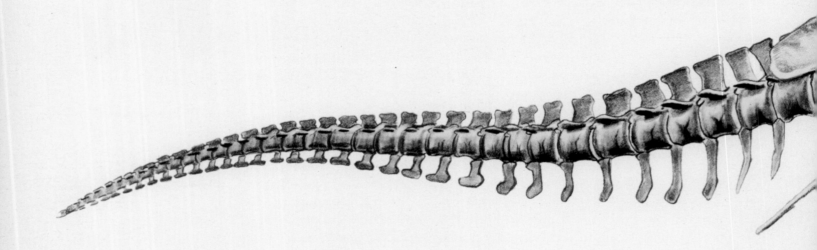

Coelurus fragilis
The one with hollow, fragile bones

This dinosaur was discovered in the 1870s from the Como Bluff site, Wyoming, USA, and described in 1879 by the famous dinosaur hunter Othniel Charles Marsh. He named it *Coelurus fragilis* because it had delicate hollow bones – an unusual feature of Late Jurassic dinosaurs of the time as most had fairly robust, heavy skeletons. *Coelurus* was about two metres in length with an elongated neck and gracile body, and weighed up to 20 kilograms. Much of the body skeleton is now known, but only the lower jawbone is known for the skull. A recent redescription of all the known *Coelurus* fossils was published with an accurate skeletal reconstruction by Dr Kenneth Carpenter and colleagues from the Denver Museum of Natural History. Scientists currently debate where *Coelurus* fits in with the evolution of the feathered dinosaurs but it is regarded by consensus as being somewhere at the base of the group containing tyrannosaurs and the more specialised forms (called maniraptorans) that make up the rest of this book.

Artist's note: This alert and opportunistic predator has been furnished with a simple insulating covering of filamentous feathers that help to prevent the loss of body heat. Found in sites with other large sauropods, *Coelurus* has been reconstructed seeking prey among the giant steps of a *Brachiosaurus*, in much the same manner as egrets do today among the big game of Africa.

Infraorder: Coelurosauria	
Family: Coeluridae	
Age: Late Jurassic	
Locality: Principally Como Bluff locality, with isolated bones from a few other sites within the Morrison Formation, Wyoming, USA.	

Tanycolagreus topwilsoni

Elongated limbs, named in honour of Mr Top Wilson

This dinosaur was collected in 1995 by the Western Paleontological Laboratories Inc. from the famous Bone Cabin locality of Wyoming. At first scientists identified the partial skeleton as being *Coelurus* but closer examination revealed it to be a new animal, named *Tanycolagreus* by Dr Kenneth Carpenter and his colleagues in 2005. *Tanycolagreus* was about four metres long and had a narrow face with a deep snout profile. A reconstruction of the complete skeleton of *Tanycolagreus* is on display the Museum of Ancient Life at Thanksgiving Point in Utah. Although relationship of *Tanycolagreus* to the other theropods is uncertain, some specialists believe that it might indeed be the earliest member of the tyrannosauroid group.

Artist's note: The hunting ground for this agile predator was the rainforest floor, an area coloured by the browns and greys of leaf litter, green mosses and by the dark shadows of trees and liana tangles. *Tanycolagreus* has a series of dark blotches and bars over base colours of orange-brown and grey. These markings allow it to blend with its surroundings and would be of most use when stalking prey – in this instance the small herbivore *Drinker nisti*, from cover.

Infraorder:	Coelurosauria
Superfamily:	Tyrannosauroidea?
Family:	Indeterminate
Age:	Late Jurassic
Locality:	Principally Bone Cabin locality, with isolated bones from a few other sites within the Morrison Formation, Wyoming, and Utah, USA.

Dilong paradoxus
The paradoxical emperor dragon

The discovery of this small predator set the worldwide media alight with the news that an early relative of *Tyrannosaurus rex* had been discovered that showed feathers were present. *Dilong* reached only 1.6 metres in length, but it shows clear affinities with the mightiest of all predators, *T. rex,* in the nature of its teeth and in the structure of its skull with a single fused snout bone (the nasal). The tail shows clearly that long filamentous fibres covered the animal's skin. *Dilong* lived in ancient China alongside a great diversity of other dinosaurs, mammals, birds, pterosaurs, reptiles and amphibians. It was one of the larger predators so far discovered in this ecosystem, although the nature of the lake deposit seem to preserve mainly smaller animals intact, so perhaps larger hunters lived near there but have not, as yet, been discovered.

Artist's note: In this painting *Dilong* is shown ambushing the gentle herbivore *Liaoceratops*. I have studied the colour patterns of similar-sized, modern predators, such as the big cats, and have given this animal a cryptic, spotted pattern-ing useful for disguise. It has short, branched, filamentous feathers which would have provided insulation.

Infraorder: Coelurosauria

Superfamily: Tyrannosauroidea

Family: Indeterminate

Age: Early Cretaceous

Locality: Lujiatun site near Beipiao city, Liaoning Province, China (Yixian Formation).

Guanlong wucai
The 'crown dragon' from the five-coloured rock

A near-complete skeleton of *Guanlong* was discovered in 2002 on the Sino-American expedition to Xinjiang Province in the far reaches of western China. Its skull had a strange backwards-facing hollow crest, possibly adorned with elaborate colours to attract mates or to warn away other males. At around three metres in length, *Guanlong* was a small but agile predator that stalked the jungles of central Asia in the shadows of slightly larger theropods such as the five-metre-long *Monolophosaurus*. It most likely preyed on small mammals, lizards and the young of other larger dinosaurs. In having U-shaped teeth at the front of its jaws, and the nasal bones of the snout fused into a single unit, it is regarded as a primitive member of the superfamily Tyrannosauroidea. Given that *Dilong*, an early member of the tyrannosauroid lineage, had feathers, there is a high probability that *Guanlong* was also covered with a similar shagreen of fibre-like feathers.

Artist's note: The distinctive nasal crest was far too fragile to be used for contact or agonistic encounters. Similar to the crests of modern currasows, it has been reconstructed with the leading edge reinforced with harder, calloused skin to afford protection when pushing through the scrub. The large openings in the bone of the crest may have been air-filled cavities, covered with a loose skin that could be inflated for display. This is similar in manner to the throat sacs of frigate birds or the elaborate lappets and horns of tragopan pheasants.

Superfamily: Tyrannosauroidea

Age: Late Jurassic

Locality: Junggar Basin, Xinjiang, China.

Eotyrannus lengi

The 'dawn tyrant' named in honour of Mr Gavin Leng

The eroding white chalk cliffs on the southern beaches of the Isle of Wight, UK, have been a treasure-trove of dinosaur discoveries for over the past 150 years. Recent finds included the partial skeleton of a small tyrannosaur, *Eotyrannus*, which was about five metres in length and weighed up to 400 kilograms. Its skull had unfused bones indicative of a juvenile, so an adult animal would have been significantly larger, perhaps as big as eight metres. The specialised pelvic structure and features of its skull and teeth show that *Eotyrannus* was part of the tyrannosaurid group, although it still retained its elongated arms and long fingers, a feature lost in later larger tyrannosaurs. The slender hind limbs also give it a more sporty appearance compared to the larger tyrannosaurs. It undoubtedly used its long arms for grasping prey, such as the one-metre-long *Hypsilophodon*, whose bones have been found in abundance from the same formation.

Artist's note: The proportions of this animal show it to be a juvenile. The sparse feathered covering and dispersal of cryptic juvenile spotting are indicative of the transitional stage to adulthood. The testing of its mettle against the formidable *Polacanthus* also indicates the inexperience of this young predator.

Infraorder: Coelurosauria

Superfamily: Tyrannosauroidea

Family: Tyrannosauridae

Age: Early Cretaceous

Locality: Isle of Wight, England (Wessex Formation).

Alioramus remotus

The other branch from a remote location

Alioramus was discovered in the remote deserts of Mongolia on the Russian–Polish expedition of the 1960s and named by Soviet dinosaur expert Dr Sergei Kurzanov in 1976. *Alioramus* is known from a fragmentary skull and bits of the body skeleton. It was about six metres long, roughly half the size of *Tyrannosaurus*, and possibly weighed up to 500 kilograms. Its most distinctive feature is the rows of knobby processes along the mid-line of its snout, almost certainly an ornamentive feature used either to attract mates, or as a deterrent to other males. *Alioramus* appears to be a close relative of the larger Asian tyrannosaur *Tarbosaurus*, according to a study of the skull characteristics of tyrannosaurids by Dr Philip Currie of the Royal Tyrrell Museum, Canada.

Artist's note: Known to science only by the skull, this head study is life-size. The rudimentary hair-like feathers can be seen around the base of the distinctive median horns.

Superfamily: Tyrannosauroidea

Family: Tyrannosauridae

Age: Late Cretaceous

Locality: Nogoon Tsav Beds, Bayankhongor Province, Mongolia.

Alectrosaurus olseni

The unmarried or single lizard, in honour of George Olsen

Alectrosaurus was originally named from just a hind limb skeleton found in Mongolia in 1923 on the American Museum's expeditions to central Asia. It was originally described by Charles Gilmore 10 years later, but further specimens including a skull and other bones were uncovered. Once thought to have long arms, this was disproven by more recent finds, showing it to be a more typical tyrannosaurid, albeit one that is poorly known. *Alectrosaurus* grew to around five metres, and possibly lived in a humid sandy environment characterised by wind-swept dunes, where it most likely hunted small ceratopsian dinosaurs such as *Protoceratops*.

Artist's note: The bold vertical barring of this *Alectrosaurus*, reminiscent of the stripes of the tiger, provides camouflage enabling it to stealthily approach its prey. An indication of feathered vestiges are apparent at the base of the forearms and upon the thighs.

Superfamily: Tyrannosauroidea?

Family: Tyrannosauridae?

Age: Late Cretaceous

Locality: Iren Dabasu Formation, China; Bayanshiree Svita (Omnogov), Mongolia.

Appalachiosaurus montgomeriensis
Appalachian lizard from Montgomery County

Appalachiosaurus is one of the most recently discovered tyrannosaurs from North America, based on a partial skull and lower jaw, the hind limbs, some vertebrae and the pelvis. It was clearly a large theropod, as the only known example is a juvenile some seven metres in length and estimated to weigh around 600 kilograms. It might be related to *Alioramus* from Asia as this dinosaur also had a row of bony crests on the snout. *Appalachiosaurus* lived on what was once an ancient island continent, called Appalachia. An unusual feature of the skeleton was that two of its tail vertebrae were fused together, suggesting healing after an injury.

Artist's note: This is a life-sized depiction of the head of a sub-adult. By this stage of its development, the animal was large enough to have dispensed with its juvenile insulating feathers. I have displayed the early stage of the growth of adult brow horns, which was another feature of its adolescence.

Superfamily: Tyrannosauroidea

Family: Tyrannosauridae

Age: Late Cretaceous (c. 77 mya)

Locality: Montgomery County, eastern Alabama, USA.

45

Tarbosaurus bataar

The alarming hero reptile

Tarbosaurus, the second-best-known tyranno-saur, was first discovered by the Russian–Polish expeditions to Mongolia in the 1950s. Closely resembling *Tyrannosaurus*, it was a large, formi-dable predator, about 14 metres in length and weighing about five tonnes. *Tarbosaurus* differs from *Tyrannosaurus* in subtle ways – certain bones of the skull contact each other in a special fashion not seen in other tyrannosaurs except for its closest relative, *Alioramus*. The nasal bone at the front of the snout has lost a special process, the lacrimal process, and there is a dominant contact between the upper jaw bone (maxilla) and the lacrimal bone of the cheek. *Tarbosaurus* was the dominant top line predator in the ancient Mongolian ecosystem between 65 and 70 million years ago. It undoubtedly hunted large plant eaters such as the hadrosaur *Saurolophus* and some of the smaller ceratopsians such as *Protoceratops*.

Artist's note: In this painting, an adult female *Tarbosaurus*, hunting prey with its offspring, has isolated a *Saurolophus* from its herd. With one bone-crunching bite to the neck, the prey is sufficiently incapacitated to enable the young to finish the kill. In comparison to the adult, the young *Tarbosaurus* still retains some of its juvenile feathered covering.

Superfamily: Tyrannosauroidea

Family: Tyrannosauridae

Age: Late Cretaceous

Locality: Nemengt Svita, Mongolia.

Tyrannosaurus rex

The king of the tyrant lizards

For many years, *Tyrannosaurus* was the largest of all known predatory dinosaurs. A relatively complete skeleton was discovered in Montana in 1902 and another soon after in 1908, which enabled a mounted skeleton to be displayed at the American Museum of Natural History. This display has inspired many children to become palaeontologists, including the late Stephen Jay Gould. *Tyrannosaurus* reached a maximum size of about 12.5 metres and weighed up to six tonnes. It took a relatively long time to reach this size – its main growth spurt occurring between the age of 14 and 19 years. From its teeth we can tell *Tyrannosaurus* was an avid carnivore, but its limb mechanics imply it was only capable of running at about 17–20 kilometres per hour, so it presumably ambushed unwary prey or attacked mainly slow moving beasts. We can guess from its close relatives that *Tyrannosaurus* probably had a downy feather covering when it was a hatchling, but we have no direct fossil evidence of it bearing feathers.

Artist's note: This fearsome individual is an adult male, as indicated by the prominent brow horns and the battle-scarrings from age-old encounters. There is staining to the teeth and to the skin around the mouth – dried blood from a recent, hapless victim. In this area there is scarring and abrasion, caused by the use of its immensely powerful jaws to smash through the bones of its prey.

Superfamily: Tyrannosauroidea

Family: Tyrannosauridae

Age: Late Cretaceous

Locality: Many localities in the USA, principally Montana, South Dakota, Colorado, Wyoming, Utah and New Mexico; also Alberta and Saskatchewan, Canada.

Daspletosaurus torosus
The frightful brawny lizard

Daspletosaurus was originally discovered and thought to be a species of *Albertosaurus*. It reached about nine metres in length and weighed up to two tonnes. Now known from some seven skulls and five partial skeletons, new finds from Montana suggest that *Daspletosaurus* is more like *Tyrannosaurus* than any other member of the tyrannosaurid family. Some palaeontologists argued that it should be placed as a separate species of *Tyrannosaurus*, but experts now separate the two forms because *Daspletosaurus* has a complex tab-like process on one of its skull bones (the postorbital bone). Dinosaur expert Dale Russell, who named *Daspletosaurus*, made the suggestion that it probably hunted the armored dinosaurs such as the horned ceratopsians, leaving its more lightly built cousin, *Gorgosaurus,* to hunt the hadrosaurs.

Artist's note: This reconstruction reveals the development of scutes and rugosites, which together with the thickened, calloused skin provided protection to the face of this formidable, bone-crushing predator. The size and pattern of these scutes correspond to the areas of greatest abrasion and wear.

Superfamily: Tyrannosauroidea

Family: Tyrannosauridae

Age: Late Cretaceous

Locality: Alberta, Canada and Montana, USA.

Albertosaurus sarcophagus
Lizard from Alberta that eats flesh

Named after the discoveries from the Red Deer River sites in Alberta, Canada, this gracile tyrannosaur grew to about nine metres in length and weighed over a tonne. *Albertosaurus* is also known from juvenile remains, enabling American paleontologist Dale Russell to reconstruct what a hatchling tyrannosaur might have looked like. The one-metre-long baby *Albertosaurus* had proportionally longer arms and legs and a more slender snout than its adult form. Although several leading dinosaur experts argue that *Albertosaurus* and *Gorgosaurus* could be the same genus, others who specialise in tyrannosaurs maintain the distinction between the two forms. *Albertosaurus* has some peculiar features in its braincase that make it distinct. This slender predator was undoubtedly faster than its contemporary, heavier kin such as *Daspletosaurus*, and was probably capable of chasing down the faster moving plant-eaters like the hadrosaurs. It bore more than 60 curved, serrated teeth in its jaws with which to tear its prey apart.

Artist's note: *Albertosaurus* is shown here with the prominent brow horns and facial scarring of an experienced and fully adult tyrannosaurid. Based on evidence of the toothmarks of *Tyrannosaurus rex* found on the hip bone of a *Triceratops*, I have depicted the carcass of the *Chasmosaurus* with lacerations to the frill and hip areas, plausibly caused by the attacking tyrannosaur.

Superfamily: Tyrannosauroidea

Family: Tyrannosauridae

Age: Late Cretaceous

Locality: Alberta, Canada and Montana, USA.

Gorgosaurus libratus
The free terrible lizard

Amongst the many discoveries from the famous Red Deer River sites in Alberta, Canada, by ace dinosaur hunter Charles H. Sternberg, was an almost complete skeleton of a lightly-built predatory dinosaur. The skull was complete although slightly crushed. The dinosaur was later named *Gorgosaurus* by Lawrence Lambe. Today at least a dozen more-or-less complete skeletons of this dinosaur have been discovered. *Gorgosaurus* reached about nine metres in length and weighed up to one tonne. It is closely related to *Albertosaurus* with which it has in the past been sometimes confused. Dinosaur experts hold that *Gorgosaurus* is a distinct genus because of its more rounded orbits for the eye, as well as having specialised features in its upper jaw not found in other tyrannosaurids. Some palaeontologists regard *Gorgosaurus* as being a scavenger, whilst others have proposed it possibly hunted the lighter-built plant-eaters such as the hadrosaurs.

Artist's note: An adult female – depicted with reduced brow horns – together with her young at the kill of a *Parasaurolophus*. The feeding posture is based upon that of the Komodo Dragon and the manner in which it processes its food. The juveniles have a cryptic patterning to disguise their presence from other predators. They also have an insulated covering of feathers which, due to their small size and mass, help to reduce the loss of heat.

Superfamily: Tyrannosauroidea

Family: Tyrannosauridae

Age: Late Cretaceous (c. 70 mya)

Locality: Alberta, Canada; also Montana and New Mexico, USA.

Pelecanimimus polyodon
The pelican mimic with many teeth

Pelecanimimus is only known from the skull and neck region, but the fossil is beautifully preserved and shows a considerable degree of fine detail that was preserved by the laminated shales of Las Hoyas, Spain. The most peculiar features of *Pelecanimimus* are a short, backwards-projecting crest on its head, and a large 'pouch' in its throat, similar to that of a pelican. It had a sharply pointed snout with many, very small needle-like teeth – a primitive feature for the group. It has been suggested that it used its teeth for both cutting and ripping – a precursor to the more advanced members of its group that evolved cutting-edged beaks. The shape of its head suggests it was probably adept at catching fish in the way modern seabirds do. It grew to about 2.5 metres in length.

Artist's note: I have given *Pelecanimimus* bold patches of pigmented skin to accentuate the display crest. The bright green of the iris is not purely ornamental – it reduces surface glare and gives *Pelecanimimus* an enhanced ability to see more fish. A dark brown or black iris reduces the amount of light reaching the posterior chamber of the eye. Eyes with light grey, blue or green pigmentation not only act as a filter, they also reflect a lot of the light back, reducing glare. These iris colours are commonly found in humans that inhabit high latitudes, where exposure to sunlight at a low angle is the norm.

Superfamily: Ornithomimosauroidea

Family: Ornithomimosauridae

Age: Early Cretaceous

Locality: Las Hoyas, Cuenca Province, Spain (c.115 mya).

Shenzhousaurus orientalis
The Shenzhou lizard from the orient

This dinosaur is known from one partial skeleton, the head bent back over the body in a death pose. It is a primitive ornithomimosaur having just a few teeth at the front of its lower jaw and none in its upper jaw. The specimen was found and collected by local Chinese farmers and much of the skeleton was lost in the process. *Shenzhousaurus* was a relatively small ornithomimosaur, about one-and-a-half to two metres in length. As with some other members of its group, its gut region contains gastroliths or stomach stones suggesting it ate a variety of foods, both small animals and plant matter.

Artist's note: Fossils located in the same deposit as *Shenzhousaurus* indicate that the habitat of this animal was one dense with mosses, ferns and the damp, shady environment was rich in amphibian life, including the discoglossid frog *Callobatrachus sanyanensis*, illustrated here. Its relatively long arms with their manipulative hands may have helped *Shenzhousaurus* in its search for food. The dark colour of its skin and feathers helped to maintain its body temperature, absorbing heat from the meagre sunlight of its shadowy world.

Superfamily: Ornithomimosauroidea

Family: Ornithomimosauridae

Age: Early Cretaceous

Locality: Sihetun site, near Beipiao, Liaoning Province, China.

Harpymimus okladnikovi

The harpy mimic named in honour of Alexey Okladnikov.

Harpymimus is known from a relatively complete skeleton lacking only bits of the hip, arms and legs. It was discovered in the early 1980s on a joint Soviet–Mongolian Expedition and named after the harpies of Greek mythology. Like *Shenzhousaurus* it retains a few teeth only on the lower jaw, but in general form resembles most other ornithomimosaurs with its long neck, well-developed beak forming the upper jaw, and long powerful legs. The neck was around 60 centimetres in length suggesting an overall body size of close to three metres. Dinosaur experts regard *Harpymimus* as a fairly primitive ornithomimosaur. The presence of its teeth in the lower jaw and beaked upper jaw suggests it used its mouth for grabbing and holding small struggling prey, such as small reptiles and mammals.

Artist's note: The long, powerful legs of these animals enabled them to flee rapidly from potential trouble. However, as with all creatures likely to become prey, it is better to avoid detection in the first place. The distinctive and prominent barring evident on this animal assists in providing camouflage, and is reminiscent of the colour scheme of many of today's forest herbivores such as the okapi, bongo or some of the duikers of Africa. The strong, vertical black and white barring served to break up the outline of an animal that is, with its long tail, conspicuously horizontal.

Superfamily: Ornithomimosauroidea

Family: Harpymimidae

Age: Early Cretaceous

Locality: Dundgovi Aimag, Eastern Gobi Province, Mongolia.

Garudimimus brevipes
The Garuda mimic with a short tail

This dinosaur was named after Garuda, the Indian monkey god who flew with bird-like wings. It was discovered from a single skeleton in southern Mongolia. An earlier study of it suggested it had a small horn on its snout but this has since been proven incorrect. The foot structure is less specialised than many other ornithomimosaurs as is the nature of its first hand digit. *Garudimimus* grew to about three metres in length. It was completely toothless and probably fed on an omnivorous diet of small animals, seeds and other plant materials.

Artist's note: *Garudimimus* was the first of the toothless ornithomimosaurs. This illustration shows the development of the pincer-like beak at the front of the skull together with a tomium-like edge to the upper and lower jaws. The tomium in the beak of birds provides a scissor-like action for processing food – a function previously performed by teeth. Note the large eye, ever alert for predators.

Superfamily: Ornithomimosauroidea

Family: Ornithomimidae

Age: Late Cretaceous

Locality: Bayanshiree Svita, southern Mongolia.

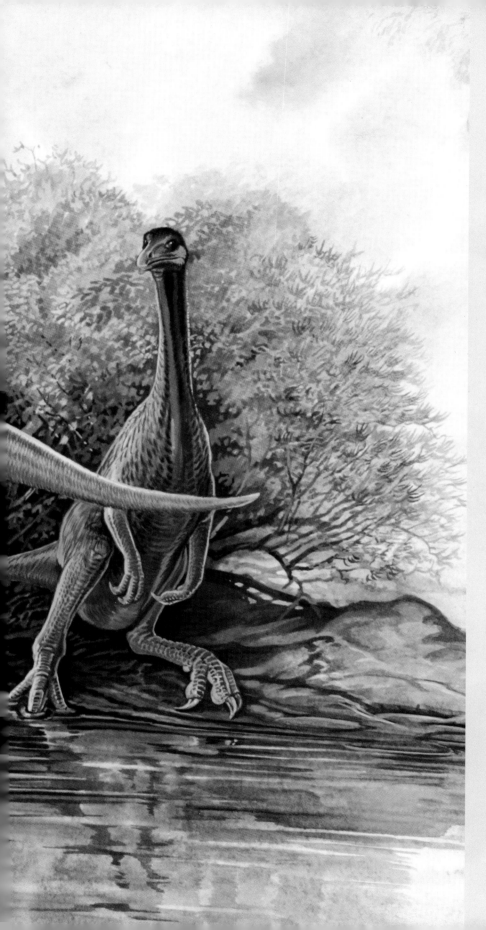

Sinornithomimus dongi

The Chinese bird mimic named after Dr Dong Zhiming

One of the best-preserved assemblages of ornitho-mimosaur skeletons ever found was uncovered in Mongolia in 1997 on a joint Japanese–Mongolian expedition. The deposit contained some 14 skel-etons (nine of which were quite complete) of a new ornithomimosaur which Dr Kobayashi and colleagues later named *Sinornithomimus*. This herd of *Sinornithomimus* comprised only two adults, a sub-adult and 11 juveniles. The largest specimen shows it reached two metres in length. The adults were better built for faster running than the juveniles – if danger approached the young were most vulnerable. The strategy for protecting the young for this species was to live gregariously so the adults could watch over the young. Gastroliths or stomach stones found with the fossil remains suggest they ate plant mate-rial that needed further grinding up in the gut. *Sinornithomimus* is regarded as a fairly primitive member of the ornithomimid group.

Artist's note: I have illustrated the adult male (the one keeping guard) with bold facial markings and a distinctive dorsal stripe. The female, wading in the foreground, is similar to the male albeit with a more subdued colouring. On the shore, a sub-adult displays its social subordination by a lack of facial markings and by the retention of some juvenile patterning. The down-covered juveniles are endowed with the colours and cryptic markings common to the chicks of today's flightless birds.

Superfamily:	Ornithomimosauroidea
Family:	Ornithomimidae
Age:	Late Cretaceous
Locality:	Alshanzuo Banner, Nei Mongol, China.

Dromiceiomimus brevitertius

The emu mimic with a short third finger

The discovery of this dinosaur goes back to the 1920s when a skeleton was uncovered in Alberta, Canada, and initially identified as belonging to *Struthiomimus*. Then in the 1970s American palaeontologist Dale Russell restudied various remains of North American ornithomimosaurs and relegated the specimen (along with other fossils) to a new form, *Dromiceiomimus*. Reaching nearly four metres in length *Dromiceiomimus* was a moderate sized ornithomimosaur, weighing around 150 kilograms. The species name comes from its relatively short third finger on the hand. Although Dale Russell originally suggested it was a carnivore feeding on small animals and other dinosaur eggs, more recent evidence of gut contents and stomach stones suggests that the group was most likely omnivorous, with a largely vegetarian diet.

Artist's note: In this family group, the adult male acts as sentinel while the female and sub-adult forage head-down among the leaf litter in the thick vegetation of riverine plains, bathed in a warm, sub-tropical climate. Although the density of this vegetation would have provided *Dromiceiomimus* with adequate cover, their ability to avoid detection has been enhanced by a cryptic pattern of white bars, reminiscent of the African bongo. Surprisingly lightly built, *Dromiceiomimus* could sprint into action at the first sign of danger.

Superfamily: Ornithomimosauroidea

Family: Ornithomimidae

Age: Late Cretaceous

Locality: Red Deer River, Alberta, Canada.

Gallimimus bullatus

The fowl mimic with blistered bone surface

Gallimimus was one of the largest of the ornithomimosaurs reaching some four to six metres in length and possibly weighing around 440 kilograms. It has been described from relatively complete skeletons of both adults and juveniles discovered on the joint Polish–Mongolian expeditions of the early 1970s. Although resembling other ornithomimids in overall shape, *Gallimimus* possessed short hands, a feature that makes it distinct from other ornithomimids. One specimen of its skull preserved a fine-combed filtering apparatus on the beak, leading American palaeontologist Mark Norell and his colleagues to suggest it may have fed like a duck, filter-feeding on small organisms in the water. It may also have used its toothless beak to crop plants at the water's edge.

Artist's note: A habitat of open woodland and temporary watercourses would have provided little cover for such a large animal and regulating its body temperature would have been critical. *Gallimimus* may have dissipated excess heat by the fluttering of its throat sac or by air movement across the naked skin areas of its inner arms and legs. A short, feathered covering protected the exposed skin from the intense rays of the sun and its light colour would have helped to reflect the heat.

Superfamily: Ornithomimosauroidea

Family: Ornithomimidae

Age: Late Cretaceous

Locality: Omnogov, Gobi Desert, Mongolia (Nemegt Formation).

Struthiomimus altus

The tall ostrich mimic

Struthiomimus was one of the first members of the ornithomimosaurs to be described from a near-complete skeleton discovered in 1914 by Barnum Brown at the Red Deer River. It is now known from two relatively complete skeletons and a small number of partial skulls and skeletons. American dinosaur expert Henry Fairfield Osborn (who described *Struthiomimus*) was the first person to make direct inferences about similar lifestyles between the ornithomimosaurs and modern ratite birds. He also noted similarity between the long three-fingered hands of *Struthiomimus* and those of the living three-toed sloth. He suggested they used their hands to grab branches to feed on leaves and seeds. Indeed the hands of *Struthiomimus* are longer than its skull or the upper arm bone. *Struthiomimus* grew to about four metres in length. Its agile lifestyle would have enabled it to live around larger predators such as *Gorgosaurus*, *Albertosaurus* and *Daspletosaurus*.

Artist's note: For this painting of *Struthiomimus* running in full flight, I drew inspiration from the ostrich. However there are significant differences. Due to the lack of a long, counterbalancing tail, an ostrich when running needs to hold its head upright and above the body. *Struthiomimus*, with its long tail and arms slung below its body, would have enjoyed a greater manouvreability with its lower centre of gravity.

Superfamily: Ornithomimosauroidea

Family: Ornithomimidae

Age: Late Cretaceous

Locality: Red Deer River, Alberta, Canada.

Ornitholestes hermanni

**The bird robber named in honour of
Mr Hermann**

Discovered in 1903 from the famous Bone Cabin
Quarry in Wyoming, *Ornitholestes* is still known
only from the one skeleton. For the many years
it has been depicted with strange horn-like
crest on the snout. However this misconcep-
tion was corrected when the skull was restudied
in detail and shown to have a misplaced nasal
bone forming the strange horn-like projection.
Ornitholestes was so-named by Henry Fairfield
Osborn as he envisaged the little dinosaur sneaking up on birds like *Archaeopteryx* which lived
at about the same time, but has not been found
in the same deposits. The species name honours
Adam Hermann who prepared the specimen.
Ornitholestes was about two metres in length,
perhaps weighing around 15 kilograms. It had
fairly long second and third digits on its hand,
a trait which makes it fairly specialised amongst
theropods. It most likely fed on the lizards, frogs,
salamanders and small mammals that lived with
it in a large river valley habitat.

Artist's note: With its long body and small head
Ornitholestes had a reasonably low centre of
gravity. It may have been more at home clamber-
ing among trees than running on the ground. It is
also credible that one of its long fingers possessed
a manipulative range, not unlike an opposable
thumb. As with many arboreal predators, I have
given *Ornitholestes* good camouflage in the form
of a series of blotches and bands over a linear
network of white.

Coeluria incertae sedis

Age: Late Jurassic

Locality: Como Bluff, Wyoming
(Morrison Formation).

Proceratosaurus bradleyi

The first horned lizard named after Mr Bradley

This enigmatic dinosaur is known only from a single incomplete skull with jaws found in 1910 near Mitchinhampton in Gloucestershire, England. It features a small crest on the snout that hinted at its being a close relative of the well-known Late Jurassic 'horned' theropod *Ceratosaurus*, although the most recent researchers place it as a primitive colelurosaurian similar to *Ornitholestes*. *Proceratosaurus* was about three metres in length and weighed around 250 kilograms. It lived in the Middle Jurassic during a time when there were few large theropods. It had a lightly-built skull with many strongly-recurved teeth, suggesting it may have been quite an agile, fast-moving predator.

Artist's note: The purpose of the prominent median crest, evident in this animal, was either for display or as an indication of sexual dimorphism. The signalling advantage of this feature would be further enhanced by its strong colourings. From the study of modern birds, it is credible to conclude that the majority of these dinosaurs also possesssed a throat sac for the purpose of temperature regulation and for display.

Coeluria incertae sedis

Age: Middle Jurassic

Locality: Mitchinhampton, Gloucestershire, England.

Nqwebasaurus thwazi

The fast running lizard from the Nqweba region,

Nicknamed 'Kirky' when it was found in 1996 in the Kirkwood Formation, South Africa, *Nqwebasaurus* was an exciting discovery as previously known primitive coelurosaurs were mainly from the northern hemisphere. This find pushed the Gondwana record of the group back another 50 million years. The partial skeleton includes much of the skull, backbone, hips and some of the arm and leg bones. *Nqwebasaurus* was only about a metre in length and had a relatively long first finger on the hand, a feature which could ally it with the compsognathid clan. This finger could oppose the second and third digits, which could well be a specialisation for grabbing its prey. Some dinosaur experts regard it as allied to either *Ornitholestes* or to the ornithomimosaurids. *Nqwebasaurus* was also preserved with gastroliths or stomach stones inside it. This suggests it ate a wide range of foods possibly including plant material such as seeds which would require further reduction in the gut by the grinding action of the stones.

Artist's note: This animal is a sub-adult which I have illustrated in transition between a spotted juvenile and an adult with bold counter-shaded blotches. These blotches, with white interspacing, provide excellent camouflage in a shady, forest environment.

Infraorder: Coelurosauria

Family: Compsognathidae?

Age: Early Cretaceous

Locality: Kirkwood, Cape Province, South Africa (Kirkwood Formation).

Compsognathus longipes
The slender-jawed one with a long tail

Compsognathus was discovered in 1850 from the Solnhofen lithographic limestones of Bavaria, Germany. The skeleton was about one metre long and was for many years known as the smallest dinosaur. Remarkably the specimen shows the remains of it last meal, a lizard, *Bavarisaurus*, inside its gut cavity, demonstrating it was a fast-moving agile predator. The late Dr Beverley Halstead suggested *Compsognathus* might have been a swimming dinosaur as it lived near the sea and was buried in a marine deposit, but the skeletal anatomy shows no specialisations to support this idea. A second complete specimen of *Compsognathus* is now known from France showing it grew to around 1.4 metres. Although the describers of the new specimen allocated it to a new species, other experts believe these differences are attributed to changes in growth and suggested that it probably belongs to the same species as the original specimen.

Artist's note: *Compsognathus* was found on the semi-arid islands of what is now Europe. Modern ground-dwelling predatory birds, found in similar habitats, are predominately lighter in colour with a pattern of small spots or bars for camouflage. An excellent example of this is the roadrunner, a bird to which I turned as a model for *Compsognathus*. I have given it an expandible throat sac, critical for temperature regulation and most useful when devouring substantial prey, as this hapless *Bavarisaurus* is discovering.

Infraorder: Coelurosauria

Family: Compsognathidae

Age: Late Jurassic

Locality: Eichstätt, Bavaria, Germany (Solnhofen shales)

Huaxignathus orientalis
The oriental jaw from old China

The fossil remains of this dinosaur were found in 2003 in a quarry by farmers from the village of Dabangou in northern China. Only two specimens are known, the best one being an almost complete skeleton – missing only parts of the tail – preserved in five large slabs. *Huaxiagnathus* grew to around 1.8 metres and represents the most primitive member of the compsognathid family because of its very long hands, amongst other features. Although the specimens do not have preserved feathers, it is most likely that *Huaxiagnathus* bore a shagreen of fine hair-like feathers as did other better-preserved compsognathids such as *Sinosauropteryx*. *Huaxiagnathus* lived in the ancient Liaoning region and probably hunted smaller food items such as frogs, lizards, salamanders and the young of other dinosaurs.

Artist's note: I have illustrated *Huaxignathus* as a sub-adult, in a transitional colour stage from a spotted juvenile to a boldly-barred adult.

Infraorder: Coelurosauria

Family: Compsognathidae

Age: Early Cretaceous

Locality: Sihetun region, near Beipiao, Liaoning Province, China (Yixian Formation).

81

Sinosauropteryx prima

The first Chinese lizard wing

Sinosauropteryx was the first dinosaur ever found with proto-feathers preserved on the body and is now known from three near-complete skeletons. A recent study alternatively suggests it had a stiff fibrous ridge on its back. The first specimen totally astounded the palaeontological world because, in addition to having hair-like feathers, there were the remains of a mammal in its gut and what were thought to be eggs in its oviduct. *Sinosauropteryx* was about 1.2 metres in length and hunted small mammals and other creatures that lived alongside the ancient Liaoning lake systems. Its body was covered in a fine down of short hair-like feathers which are now known to be the first stage of development in longer, more complex feathers that occur on dinosaurs and birds. It was very much like other compsognathids differing only in the proportions of its skull and limb bones.

Artist's note: The exquisite fossil of this dinosaur leaves no question about the presence of feathers on these small dinosaurs. Even the faint outline of coloured barring is evident along its tail. As this is a female, I have avoided strong ornament and colour around the head and throat. The large orbit for the eye and the presence of a recently ingested mammal in the gut indicates an animal that probably hunted in the shadows or at night. A high contrast barring of black and white helps to disguise the body outline in such conditions.

Infraorder: Coelurosauria

Family: Compsognathidae

Age: Early Cretaceous

Locality: Sihetun region, near Beipiao, Liaoning Province, China (Yixian Formation).

Juravenator starki

The hunter from the Jura mountains named in honour of Mr Stark

Nicknamed 'borsti' because it might have had fine hairs like a hunting dog, the complete fossil of *Juravenator* showed the remains of a scaly type of skin cover. It was discovered by fossil hunter Klaus-Dieter Weiss in a lime pit close to Eichstätt, Germany, where *Archaeopteryx* and *Comspognathus* were found. At around 80 centimetres in length, it was a small dinosaur, possibly a juvenile, comparable in size to other compsognathids. *Juravenator* shows that primitive coelurosaurians did not all have feathers, or that the feathers were not yet developed at the particular growth stage of this individual. An alternative explanation is that this little dinosaur is much more primitive than first thought, and represents a primitive maniraptoran, a group that is known not to have developed feathers. *Juravenator* lived close to the sea and probably hunted small animals near the water's edge.

Artist's note: Because its fossil showed no evidence of feathers, I have illustrated *Juravenator* accordingly. However, it is possible that this young animal may either have been predominantly naked or covered in a downy fluff similar to the chicks of parrots – in which case, such fine feathers may not have survived the rigours of fossilisation.

Infraorder: Coelurosauria

Family: Compsognathidae?

Age: Late Jurassic

Locality: Quarry Stark, west of Schamhaupten, Bavaria, Germany.

Sinocalliopteryx gigas
The giant Chinese beautiful feather

Sinocalliopteryx is not only the biggest compsognathid but also the largest known dinosaur fossil with feathers attached to it. Most surprisingly, the fossil skeleton of this dinosaur was found with the leg of a small dromaeosaur inside its gut, indicating that compsognathids could also be fierce hunters rather than just foragers on small lizards and mammals. Reaching two-and-a-half metres in length, *Sinocalliopteryx* had relatively long hands in relation to its arms, which were longer than in other compsognathids. Four stones found inside the gut region of the fossil are interpreted as being gastroliths, or gizzard stones, that helped it digest its food. It had feathers all over its body, the longest ones occurring on its hips, thighs and the base of its tail. It also had foot feathers – a feature not restricted to just the advanced dromaeosaurids and early birds, but arising much further back in theropod evolution.

Artist's note: The leg of a dromaeosaur ingested by *Sinocalliopteryx* indicates that it was capable of bolting down food of a substantial size. To accommodate this I have endowed it with a bright orange, expandable throat sac. The strong pattern of black and reddish-brown patches upon white give the animal a cryptic advantage when stalking its prey. In this instance I have reconstructed a non-specific dromaeosaur, estimating the size of the animal from the known leg elements.

Infraorder: Coelurosauria

Family: Compsognathidae

Age: Early Cretaceous

Locality: Liaoning Province, China.

Alxasaurus elesitaiensis
Alxa desert lizard found at Elesitai village

Alxasaurus is known from five skeletons recovered from the Alxa Desert region of inner Mongolia, collected in 1988 on the joint Sino–Canadian dinosaur expeditions. It was an exciting discovery as previously very little was known of the strange therizinosaur group. Most of the body skeleton of *Alxasaurus* is preserved but only the lower jaw is preserved from the skull. The largest specimen reached 3.8 metres in length. Like other therizinosaurs, it had a long neck, large arms with enlarged hand claws, a robust pelvis and a short tail. *Alxasaurus* is characterised by the primitive nature of its many teeth – it had 40 on its lower jaw. More evolved therizinosaurs had fewer teeth.

Artist's note: This painting depicts two males fighting over territory. The older and larger male, on the right, has a face flushed with red, which signals his status to the younger, subordinate interloper.

Infraorder: Coelurosauria

Family: Therizinosauridae

Age: Early Cretaceous

Locality: near Elesitai, Alxa Desert region, Inner Mongolia, China.

Falcarius utahensis
The sickle-maker from Utah

A newly discovered treasure-trove of dinosaur bones found in Grand County, Utah, USA, contains maybe hundreds of *Falcarius* skeletons. An initial description of this new dinosaur was based on about ninety per cent of the recovered skeleton when it was published. *Falcarius* was about four metres in length and is possibly the most primitive member of the therizinosaur family, mainly because it lacks features of the forearm and pelvis seen in more advanced forms. It had peculiar, enlarged front lower jaw teeth similar to those seen on the upper jaw of the buck-toothed *Incisivosaurus*. Dr Scott Sampson, one of its discoverers, described *Falcarius* as '… the missing link between predatory dinosaurs and the bizarre plant-eating therizinosaurs'. *Falcarius* reveals how the teeth and jaws and gut region were the first parts of the skeleton to change when the therizinosaur family gradually evolved from being meat-eaters to a more specialised vegetarian diet.

Artist's note: Most modern herding animals indicate sexual, social or age separation with some form of visual signaling involving colour, pattern or adornment. Of the five animals illustrated here, three are females with subdued colouration, and two are sparring males with bolder markings and erectile crests. The carbonate-rich sediment in which the fossils were found suggests the spring-fed marshy habitat evident in this painting.

Infraorder: Coelurosauria

Family: Therizinosauridae

Age: Early Cretaceous

Locality: Crystal Geiser Quarry, Grand County, Utah, USA.

Beipiaosaurus inexpectus
The Beipiao lizard with surprising features

This dinosaur is known only from a partially articulated skeleton found near Beipiao city in Liaoning Province, China. It demonstrated that therizinosaurs had feathers, signalling that feathers may have been present in a much wider range of theropods than previously thought. The unusual skeletal anatomy of *Beipiaosaurus* places it clearly within the therizinosaurs, yet it had feet somewhat like the gargantuan sauropod dinosaurs. It has a large skull compared to other therizinosaurs and strange bulbous teeth. There are feathers on various parts of its skeleton, including filamentous kinds on its forearms and legs. *Beipiaosaurus* is regarded as one of the most primitive members of the therizinosaur family, slightly more advanced than *Falcarius* because of its specialised hip and jaw features.

Artist's note: Evident in the fossil of this early therizinosaur are feathered filaments which provide a tantalising glimpse into the outward appearance of this exotic dinosaur. These hair-like filaments appear densely packed, branched on the end and hollow. Birds with similar hollow spaces in their feather barbs – as in some of the cotingas – appear white, blue or bluish-green in colour. For this reason, I have used blue and white as the predominant colours on this animal, with some black separation barring as contrast.

Infraorder: Coelurosauria

Family: Therizinosauridae

Age: Early Cretaceous

Locality: Sihetun region, Beipiao, Liaoning Province, China.

Nothronychus mckinleyi
Sloth-like claw named in honour of McKinley

Nothronychus was discovered in 2001 from the region between Arizona and New Mexico known as the Zuni Basin. It was the first definite find of a therizinosaur in North America. The fossil remains of two individuals comprised enough of the skull and body skeleton to build a fair picture of what this dinosaur looked like when alive. *Nothronychus* was between five to six metres in length, standing up to 3.6 metres tall. It had a stout, robust belly, a short tail, a small head and very long arms with 10-centimetre, curved claws on its elongated hands. Its teeth were leaf-shaped, as found in other plant-eating dinosaurs. *Nothronychus* lived in a tropical jungle environment and probably fed on a variety of plants.

Artist's note: The larger an animal becomes, the greater is its need to dissipate excess body heat. This bulky, waddling, pot-bellied, plant-eating creature was no exception. One way of accomplishing this is through the movement of air over areas of bare skin that are rich in blood capillaries, such as in the open mouth of a basking crocodile or the fluttering throat sac of many birds. However, neither of these means suited a large vegetarian animal which spent most of its waking hours plucking and swallowing vast quantities of leaves. More probably its inner arms, chest and lower neck were free of feathers. Such areas would be exposed to a cooling breeze when feeding upright, and shaded when in its horizontal resting position.

Infraorder: Coelurosauria

Family: Therizinosauridae

Age: Early Cretaceous

Locality: Southern Zuni Basin, New Mexico, USA, (Moreno Hill Formation).

Erlikosaurus andrewsi

Erlik's lizard named in honour of Roy Chapman Andrews

In Mongolian mythology, Erlik was the King of the Dead, so the name aptly applies to this ancient creature of the Mongolian desert, *Erlikosaurus*, which was discovered on the joint Soviet–Mongolian expeditions of the early 1980s. The most complete skull of any therizinosaur yet found, it featured jaws with many small peg-like teeth, an elongated snout, and a narrow but large slit-like nasal opening. *Erlikosaurus* grew to six metres in length. It was discovered and first described by Mongolian palaeontologist Perle Altangerel, who later re-examined the skull, working with US palaeontologists James Clark and Mark Norell. It was first suggested that *Erlikosaurus* ate fish, but most palaeontologists have swung around to the view that because of their large gut and similar forelimbs to modern sloths, therizinosaurs probably fed primarily on vegetation which they gathered using their large scythe-like claws.

Artist's note: The elongated snout together with rows of small, flattened teeth, indicate that *Erlikosaurus* was a plant eater, well equipped for plucking and raking-in leaves. I have given him thick, calloused skin on the face and a raised area around the nostril to protect against abrasion by twigs. The eye too, is similarly protected by a prominent brow and coarse bristles. The bluish tinge to the tongue is a feature of tree browsing herbivores, such as giraffes, who often have a high level of melanin in the tongue as a protection from sun exposure.

Infraorder: Coelurosauria

Family: Therizinosauridae

Age: Late Cretaceous

Locality: Omnogov, Baynshirenskaya Svita, Mongolia.

Segnosaurus galbinensis
The slow lizard from the Galbin region

Although it still remains a very poorly known, *Segnosaurus* was one of the first therizinosaurs to be discovered. In the late 1970s, the joint Soviet–Mongolian expeditions to the Gobi Desert found three specimens that consisted of bits of the body skeleton and a lower jaw. The jaw had 24 sharply pointed but flattened teeth indicating *Segnosaurus* may have been omnivorous, occasionally eating small lizards or insects. The shape of the lower jaw of *Segnosaurus* indicates it lacked the well-developed cheeks of other more advanced therizinosaurs. *Segnosaurus* was one of the larger members of its group, growing to about six metres in length, and may have stood about 2.5 metres high.

Artist's note: Threatened by an approaching *Alectrosaurus*, *Segnosaurus* is reconstructed rearing into a defensive position. As the larger therizinosaurs were not built for speed, it is probable that they used their enormous sickle-like claws for defence.

Infraorder: Coelurosauria

Family: Therizinosauridae

Age: Late Cretaceous

Locality: Dornagov and Omnogov, Baynshirenskaya Svita, Mongolia.

Neimongosaurus yangi

The Inner Mongolian lizard named after Dr Yang

The scant remains of *Neimongosaurus* were discovered in 1999 near Sanhangobi, Inner Mongolia, by a Chinese team of palaeontologists led by Dr Xiao-Hong Zhang and Xu Xing. It is known from two specimens comprising much of the vertebral column, ribs, skull, and parts of the pelvis and limbs. *Neimongosaurus* was of moderate size, reaching about two-and-a-half metres in length. It had an unusually long neck and strange spine – unlike those of other therizinosaurs – in that the front vertebrae of the tail have a deep pit underneath them. The lower jaw is marked by its strong curvature downwards at the front of the mouth. The teeth – with many small ridges on them similar to some plant-eating ornithischian dinosaurs – clearly suggest that *Neimongosaurus* was a plant-eater.

Artist's note: I have reconstructed this smaller member of the Therizinosauridae as a male sporting courtship colours. Although painted running in a horizontal position, the natural inclination for these animals would have been to stand upright. In this position, the most prominent areas for sexual display would be the underside of the neck and throat, the upper body and the forearms.

Infraorder: Coelurosauria

Family: Therizinosauridae

Age: Late Cretaceous (c. 92 mya).

Locality: near Elesitai, Alxa Desert region, Inner Mongolia, China.

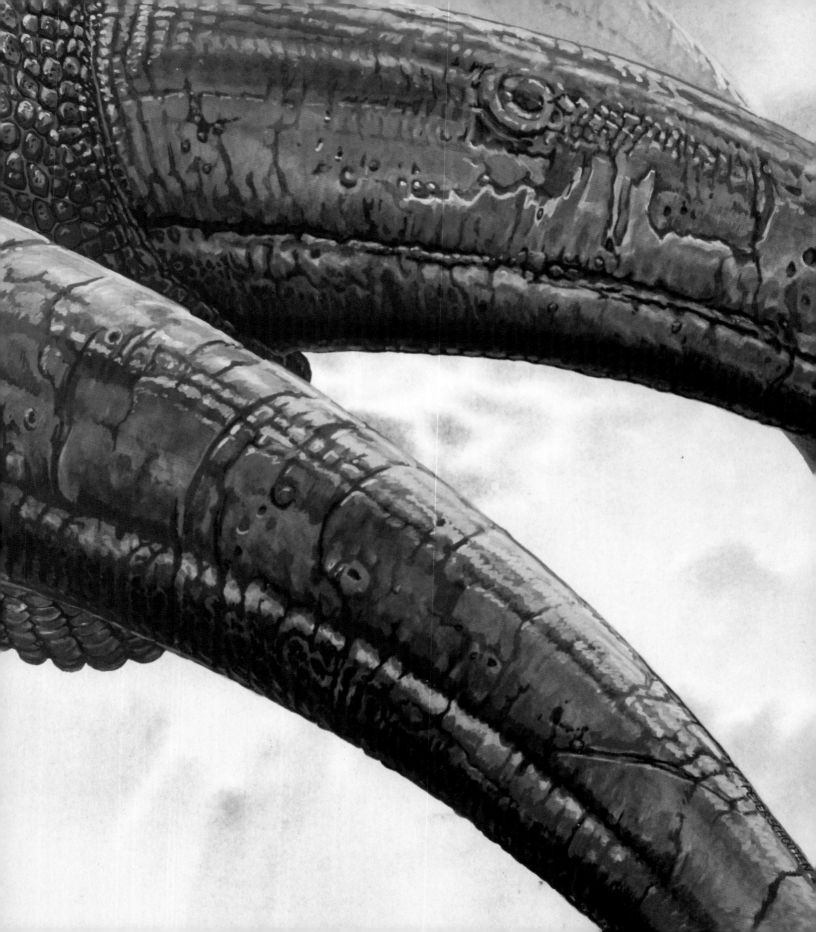

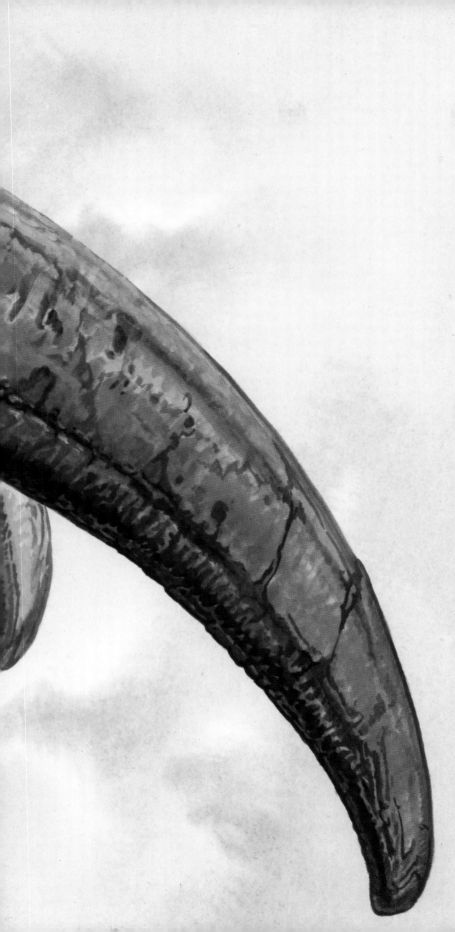

Therizinosaurus cheloniformis

The slashing claw lizard resembling a turtle

This is the largest yet most enigmatic of all therizinosaurs. The huge, curved claws of this beast, measuring nearly a metre long, were uncovered in the Gobi Desert of Mongolia in the late 1940s by a joint Soviet–Mongolian expedition. At first Russian palaeontologist Evgeny Maleev thought they belonged to a giant turtle. Later finds unearthed more of the skeleton including parts of its lower jaw and forelimbs and it was then realized that *Therizinosaurus* was actually a dinosaur. The front of the lower jaw is toothless suggesting it had a beak for cropping plants. Each of its powerful arms had three gigantic claws. Growing maybe as long as 10 or 12 metres, the giant may have been a termite eater, or perhaps just used its huge claws to slash vegetation down from trees in order to feed.

Artist's note: Because *Therizinosaurus* in known to us only by the fossilised remains of a partial forelimb, which included the hand and giant claws, there is unfortunately not enough information to enable the reconstruction of the entire animal. Its claws (shown here nearly life-size) are uniquely therizinosaurid in character, and are the largest claws of any dinosaur found to date.

Infraorder: Coelurosauria

Family: Therizinosauridae

Age: Early Cretaceous

Locality: White Beds of Hermiin Tsav, Omnogov, Mongolia.

Incisivosaurus gauthieri

Incisor-bearing lizard in honour of Jacques Gauthier

This strange little dinosaur was discovered in 2002 in Liaoning Province, China, and named because of its most peculiar buck-teeth at the front of the skull. At around one metre in length, it probably dug with its front teeth for roots and tubers along the shores of the ancient lake system. Its strange front teeth show a typical wear pattern commonly seen in various plant-eating dinosaurs. The long skull with teeth show that *Incisivosaurus* was a very primitive member of the oviraptorosaur group, as most of them have shortened heads and have lost their teeth. Some scientists have suggested its affinity could lie with the bizarre therizinosaurs.

Artist's note: Similar to the modern day Kiwi, this buck-toothed little dinosaur has an insulating covering of simple, bristly feathers. As *Incisivosaurus* was closely related to *Protarchaeopteryx,* a relatively conservative colour and pattern has been maintained. The deep indigo areas of the face and neck could have been used for display. Flushed with blood, the indigo blue would convert to tones of brilliant violet, magenta and purple.

Infraorder: Oviraptosauria

Family: Indeterminate

Age: Early Cretaceous

Locality: Sihetun region, Beipiao, Liaoning Province, China (Yixian Formation).

Avimimus portentosus

The amazing bird mimic

Originally described by Russian palaeontologist Sergei Kurzanov from a few bones and a partial skull found in Mongolia, *Avimimus* was at once thought to be so close to birds that it perhaps wasn't a dinosaur. Today it is classified within the oviraptorosaurs as one of its most primitive members. Like *Inciscivosaurus* it still retains some teeth on the front of its upper jaw. *Avimimus* grew to about 1.5 metres in length, standing about 70 centimetres high at the hips. It had a slender neck containing 14 vertebrae that lacked cavities for air-sacs, a feature seen in more bird-like dinosaurs. It had very short arms with the bones of its hand and arm fused in places, much like a modern bird. Kurzanov thought the deep groove on the forelimb bone was an attachment point for feathers. This idea has now been supported by the recent discoveries of feathered oviraptorosaur dinosaurs from China. *Avimimus* had long, slender legs, suggesting it was a fast runner. We can only guess that dinosaurs such as *Avimimus* were hunters of insects and lizards.

Artist's note: In this reconstruction, I have highlighted the potential use of the feathered forelimbs for display and was influenced by the argus and peacock pheasants of South-East Asia, which use their 'eye-spots' to capture the attention of the female, directing her eye toward the heavily ornamented head and neck. The iridescent feathering on *Avimimus* would have provided such a focal point.

Infraorder: Oviraptosauria

Family: Oviraptoridae

Age: Late Cretaceous (c. 75 mya)

Locality: Djadokhta Svita, south-western Mongolia (Nemegt Formation); also Inner Mongolia, China.

Caudipteryx zoui
Winged-tail named in honour of Zou Jihua

The spectacular discovery in 1998 of the peacock-sized *Caudipteryx* demonstrated for the first time that a dinosaur could have well-developed, branching feathers, like birds, but not be flying animals. The almost-complete skeleton adorned with feathers from Liaoning Province, China, bore the distinctive feature of a tail with a fan of large feathers radiating out from halfway along its length. Its short arms also bore well-developed feathers. Like that of other oviraptorosaurs, the skull of *Caudipteryx* is short, and shows a primitive state of having a few teeth in its jaws. Whereas most dinosaur experts regard *Caudipteryx* as an oviraptorosaur, some specialists hold that it could be a highly specialised flightless bird. The presence of stomach stones inside the gut suggests it ate plant material such as seeds. The well-developed feather fans on the arms and tail were most likely used for mating displays and no doubt bore impressive patterns and colours to attract a mate.

Artist's note: I have reconstructed *Caudipteryx* in the act of courtship. The male is displaying his semaphore-like forelimb feathers as the female is signalling her readiness to mate with her raised tail. Male birds that adopt this form of display generally sport vibrant colours designed to stand out against the shadowy backdrop. Additionally they may use reflective or refractive feathers that scatter sunlight in a scintillating way. With its feathers unfurled and throat flap and wattles engorged, I have attempted to capture *Caudipteryx* in such a moment of intense display.

Infraorder: Oviraptosauria

Family: Caudipterygidae

Age: Early Cretaceous

Locality: Sihetun region, Beipiao, Liaoning Province, China.

Protarchaeopteryx robusta
The robust first ancient wing

Protarchaeopteryx is known from a relatively complete specimen, although not all the bones of its backbone and skull are preserved. Most noticeably the specimen displays symmetrical feathers on its arms that show advanced branching patterns such as seen in modern birds. Long, branched feathers that were found associated with the main fossil could be tail feathers similar to those of *Caudipteryx*. It had a few teeth at the front of the upper jaw only and in this feature resembles *Incisivosaurus* (a closely related form which some scientist think could even be the same genus as *Protarchaeopteryx*). *Protarchaeopteryx* was about 60 centimetres in length and possibly fed on small insects and plants around the Liaoning lake system.

Artist's note: The colouring of *Protarchaeopteryx* is somewhat sombre and restrained. This, with a lack of patterning, provided the means to merge unseen into its shadowy forest environment. However, vibrant patches of colour under the forearm and tail could either be hidden, when seeking camouflage, or strikingly revealed when in display.

Infraorder: Oviraptosauria

Family: Caudipterygidae

Age: Early Cretaceous

Locality: Sihetun region, Beipiao, Liaoning Province, China.

Chirostenotes pergracilis
The narrow-handed gracile one

The first parts of this dinosaur found in Alberta, Canada, were the remains of its hands, uncovered in 1924, followed soon after by its feet, in 1932. Since then more fossil remains have been found from sites in Canada and the USA, including a partial skull and parts of the body skeleton that were uncovered in 1988 after being stored since 1923. From these remains we can determine that *Chirostenotes* was about two-and-a-half metres in length, and weighed around 40 kilograms. It was a toothless form of oviraptorosaur with very short and powerful jaws. It may well have foraged in the undergrowth searching for seeds or nuts.

Artist's note: The diet of this curious group of dinosaurs remains somewhat of a mystery. With a toothless jaw and a powerful parrot-like beak, they had evolved the tools for processing very hard food items. In this reconstruction the ground is littered with cracked and fallen nuts, although whether *Chirostenotes* actually consumed these nuts is a matter of some debate.

Infraorder: Oviraptosauria

Family: Caenagnathidae

Age: Late Cretaceous

Locality: Alberta, Canada; also Montana and South Dakota, USA.

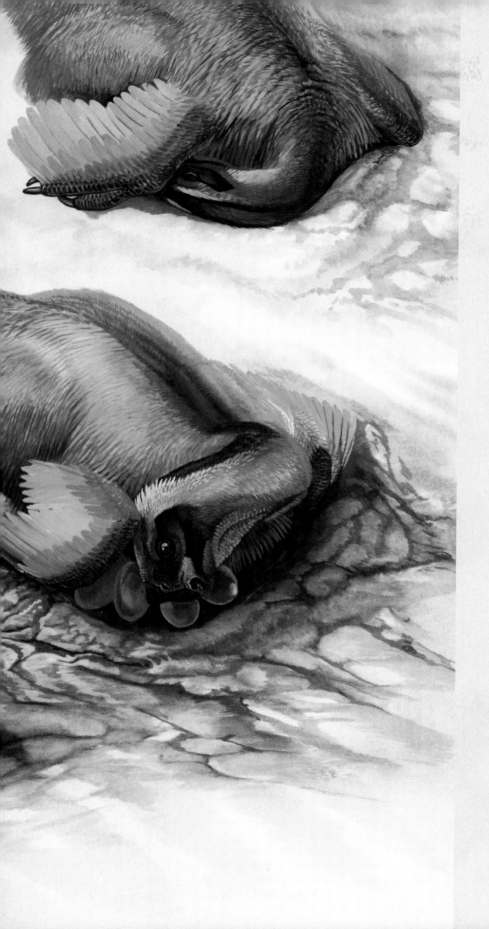

Citipati osmolskae

Funeral pyre lord named in honour of Dr Osmolska

One of the most spectacular oviraptorosaur discoveries was made in the 1990s by the joint American–Mongolian expedition to the Gobi Desert. It uncovered several complete skeletons of dinosaurs that were sitting atop of their own eggs. The dinosaurs must have come to a sudden end when wet sand dunes collapsed on top of them, preserving them on their nests. *Citipati* was about three metres in length, one of the largest oviraptorids. It had a well-developed crest at the front of its head, and its mouth was a toothless beak. Its eggs were up to 18 centimetres long, and were laid in clutches of about 22 eggs arranged in a circle. The mother dinosaur sat with legs folded underneath and arms spread out over the nest, in the same pose used by birds when brooding their eggs. Some of the eggs have embryonic dinosaur skeletons inside them.

Artist's note: This reconstruction is based exactly upon the postion of these animals as found in the fossil deposit. As a dweller of a semi-arid environment, *Citipati* probably required a light, heat-reflective overall colouration with bare patches on the face and throat for temperature regulation. Panting and fluttering of the throat sac would also have assisted in the dissipation of excess body heat. The forearms, whether raised or lowered, provided shade and allowed air movement over the eggs, thus maintaining a correct incubation temperature.

Infraorder: Oviraptosauria

Family: Oviraptoridae

Age: Late Cretaceous

Locality: Ukhaa Tolgod, south-western Mongolia (Nemegt Formation).

Rinchenia mongoliensis
Rinchen Barsbold's one from Mongolia

Although he proposed this dinosaur's name as a joke in honour of himself, Rinchen Barsbold never expected it to become formally accepted. *Rinchenia* is based on a skull and fragmentary skeleton found in Mongolia, and was about two-and-a-half metres in length. Barsbold noticed how the skull had a more well-developed crest than any previously known oviraptorosaur, giving it a distinctly cassowary-like appearance. Furthermore the crest of *Rinchenia* incorporated several skull bones not usually built into the crest of other dinosaurs. It has been suggested that its deep head and powerful toothless jaws may have made it adept at cracking clams for food.

Artist's note: In this reconstruction, *Rinchenia* is prising open a freshwater mussel. Its short, high skull, deep mandibles and presence of a curious tooth-like process projecting from the palate, indicate that this animal was capable of crushing very hard food. Positioned mid-palate, where the greatest force could be exerted, these seemingly insignificant little bumps would have been capable of breaking open objects with the force of a log splitter. With such a built-in nutcracker, it was no doubt capable of breaking open a clam with as much ease as a walnut.

Infraorder: Oviraptosauria

Family: Oviraptoridae

Age: Late Cretaceous

Locality: Omnogov, south-western Mongolia (Nemegt Formation).

Heyuannia huangi
The Heyuan one named after Mr Huang

Heyuannia was only discovered and described in 2002, based on five individual specimens, comprising most of the skeleton. It was a fairly large beast, reaching three metres in length, with fairly short arms and stout finger digits, although the first finger was very short. The skull is only partially preserved but indicates *Heyuannia* did not have a crest on its head, and the lower jaw is short and deep as in other oviraptorosaurs. In one specimen the remains of reproductive organs were thought to be preserved. The detailed description of the articulated forelimb of *Heyaunnia* by Junchang Lu and colleagues lead them to suggest that oviratorosaurs were actually secondarily flightless birds, a view that is not supported by the majority of current dinosaur specialists.

Artist's note: As a forest dweller in need of camouflage, *Heyuannia* may have had an overall cryptic colour and pattern. Like other oviraptors, its principal area for display was its head region and consequently this animal has been endowed with a dark erectile crest and a bright orange cheek patch over a white malar stripe. To a potential suitor, the naked white skin surrounding the nostrils and the eye would appear as an ultraviolet blue (when viewed through the visual spectrum of a bird) thus highlighting the brilliant orange of the iris.

Infraorder: Oviraptosauria

Family: Oviraptoridae

Age: Late Cretaceous

Locality: Guandong Province, southern China.

Ingenia yanshini
The Ingeni one named after Mr Yanshin

Ingenia was discovered in the Ingeni-Khobur Depression in the Gobi Desert on the joint Soviet–Mongolian expeditions of the late 1970s. Its fossil remains included a partial skeleton with the skull, together with the body skeletons of five other individuals. It was a small creature, reaching less than two metres in length and weighing around 25 kilograms. It had a very low crest on its head, short stout arms with robust fingers, and a very strong ankle region, possibly indicating it liked to dig for food and use its fingers to scratch around. From associated dinosaur nests we know that *Ingenia* laid a clutch of 24 eggs in a ring. It was most likely an omnivore, feeding on plants and small insects. Analyses of its skeletal features suggest that *Ingenia* may well be the most advanced of all the oviraptorosaurs.

Artist's note: There is the possibility that *Ingenia* may have sought food from lake shores or tidal mud flats, much as wading birds do today. If so, the long exposure to the intense sunlight of such foraging would require it to possess some form of heat reflective capability, such as a covering of white feathers.

Infraorder: Oviraptosauria

Family: Oviraptoridae

Age: Late Cretaceous

Locality: Hermiin Tsav, Omnogov, Gobi Desert, Mongolia.

Conchoraptor gracilis
The slender conch thief

The first discovery of *Conchoraptor* in the Mongolian Gobi Desert was thought to be a juvenile *Oviraptor*, as it lacked the well-developed crest on the head and was a small individual, about one-and-a-half metres in length. Since then several skeletons have been discovered showing it to be a small, crestless form characterised by unusually slender hands with weakly curved claws, somewhat intermediate in form between those of *Ingenia* and *Oviraptor*. *Conchoraptor* had large air chambers in its nostrils. It has been suggested that *Conchoraptor* might have been a mollusc feeder, cracking clams and other shells with its short, powerful lower jaws. *Conchoraptor* is believed to be closely related to both *Ingenia* and *Khaan*.

Artist's note: For this illustration of a male brooding eggs, I have used the modern-day parrot as a model. With the bare facial skin of a cockatoo and the bright blues and greens of the macaws, I have added a bright blue iris colour as protection from the intense sunlight and glare of its semi-arid habitat. As with other related oviraptorosaurs, *Conchoraptor* is crestless. Consequently the facial region should be the area of principal display and signalling. Note, too, the well-developed forearm feathers, spread to protect the eggs.

Infraorder: Oviraptosauria

Family: Oviraptoridae

Age: Late Cretaceous

Locality: Omnogov, Gobi Desert, Mongolia (Nemegt Formation).

Khaan mckennai
The lord named after Dr McKenna

Three skeletons of the dinosaur *Khaan* were discovered in 2000 on a joint expedition of the American Museum of Natural History and the Mongolian Academy of Sciences. Looking rather like *Conchoraptor*, *Khaan* is distinguished primarily by its differently shaped nostril and the sharper angle where the head meets the neck vertebrae. *Khaan* was about one-and-a-half to two metres in length.

Artist's note: The extensive feathering on the forelimbs of the oviraptorosaurs, displayed so prominently in this reconstruction of *Khaan*, served several purposes. They were useful for not only brooding a clutch of eggs but may also have played a role in the stability of the animal when in motion. For example, when running at speed the distended arms may have helped lower the centre of gravity, and, with feathers spread, assisted in fast directional change. Fully opened, they may also have provided a little lift, giving this animal an added advantage when outrunning a predator.

Infraorder: Oviraptosauria

Family: Oviraptoridae

Age: Late Cretaceous

Locality: Omnogov, Djadokhta Formation, south-western Mongolia.

Nemegtomaia barsboldi

The mother of the Nemegt, in honour of Rinchen Barsbold.

Nemegtomaia was a typical medium-sized oviraptorid, up to two metres in length, with a short, beaked mouth and an unusual high crest above the skull, the leading margin of which is almost vertical. Like other members of its group that have well-developed branching feathers on the arms and tail, *Nemegtomaia* almost certainly had a very bird-like appearance. The most distinctive features of this dinosaur, which separate it from the several other similar oviraptorosaurs, lie in the shape of its skull. The pre-frontal bone is absent from the skull, and its jaws have a strange kind of articulation, perhaps suggesting it had a specialised feeding mechanism. Originally this dinosaur was named from near-complete remains including a good skull as 'Nemegtia', but the name was found to be already in use by an ostracod (bivalved arthropod), so it was duly renamed *Nemegtomaia* in 2005.

Artist's note: The feature of this reconstruction is the oviraptorid crest in display. The bright red, inflated throat sac, being part of the visual display, may have also contributed to courtship vocalisations.

Infraorder: Oviraptosauria

Family: Oviraptoridae

Age: Late Cretaceous

Locality: Nemegt Formation, south-western Mongolia.

Zamyn Khondt oviraptorine

A complete skull of a new species or genus of oviraptorine dinosaur was discovered in 1979 on the joint Soviet–Mongolian expedition to the Gobi Desert at the Zamyn Khondt locality. The skull shows a well-developed crest similar in shape and form to that of *Citipati*, to which it is probably very closely related. Like *Citipati* it was a moderate-sized dinosaur, about two-and-a-half to three metres in length. We know little about this as yet unnamed beast. It probably ate hard-shelled food as it had a stout tooth-like process on its palate that would have helped crack hard seeds or shells.

Artist's note: In a similar fashion to *Khaan*, this animal has been illustrated running at speed, with the forearms deployed as a motion stabiliser. The facial display feathers include contrasting, reflective cheek stripes and an erectile crest.

Infraorder: Oviraptosauria

Family: Oviraptoridae

Age: Late Cretaceous

Locality: Zamyn Khondt, Omnogov, south-western Mongolia.

Buitreraptor gonzalesorummangas

The vulture roost robber named after the Gonzales brothers

The fossil remains of this turkey-sized dinosaur were found by two brothers, Fabian and Jorges Gonzales, in northern Argentina at the same site where the gargantuan predator *Gigonotosaurus* had been found. It was an exciting find as it represents the oldest member of the dromaeosaurid family known from South America, and it helps us understand the early evolution of the group. *Buitreraptor* is a distinctive looking dromaeosaurid, characterised by its slender, elongated snout with many small teeth that are widely spaced and lack the serrations seen in other raptor teeth. Its long legs and arms, coupled with a powerful shoulder region, imply it was a fast runner that could reach out and grasp agile small prey animals, such as frogs and small lizards. It has been suggested it lived a similar lifestyle to the modern secretary birds.

Artist's note: This very bird-like predator has been illustrated with mouth agape to emphasise its exceptionally long, slender snout and lower jaw. *Buitreraptor* was armed with a battery of tiny teeth, not dissimilar to those of the ornithomimosaur, *Pelecanimimus*. These teeth were well suited for gripping slippery prey, such as amphibians, and it is not unreasonable to assume that *Buitreraptor*, like *Pelecanimimus*, sourced its food from marshy environs.

Infraorder: Coelurosauria

Family: Dromaeosauridae

SubFamily: Unenlagiinae

Age: Middle Cretaceous

Locality: Argentina.

131

Microraptor gui

Little robber named in honour of Mr Gu Zhiwei

One of the most spectacular fossil discoveries announced in 2003 was the four-winged dinosaur *Microraptor gui,* from the famous Liaoning sites of north-eastern China. At only 77 centimetres in length, this dinosaur is one of the smallest adult species of theropod known. It had well-developed pennaceous flight feathers on both its arms and legs, along with a typical reptilian tail that also sported a fan of broad feathers. The feathered arms and legs may have helped it to glide down from the trees but its legs could not have been employed in flapping motions because its hip joints had limited flexibility. Perhaps it folded its feathered legs underneath itself to form a biplane arrangement when gliding. *Microraptor gui* was clearly an opportunistic predator that chased insects and other small prey, and used its gliding abilities to avoid being caught by the larger predatory dinosaurs.

Artist's note: Here *Microraptor* is in mid-glide, with forewings widely spread and hind limbs drawn underneath its body. The asymmetrical flight feathers provide lift, while the contour feathers of the wings and tail provide stability and pitch control. The edges and tips of the flight feathers are strengthened by a high level of the black pigment, melanin. Sexual display features are visible in the brightly coloured throat sac, cheek patch and feathered crest. The mottled black, brown and white colours enabled *Microraptor* to blend with the moss- and lichen-covered branches of its canopy home.

Infraorder: Coelurosauria

Family: Dromaeosauridae

Age: Early Cretaceous

Locality: Sihetun region, Beipiao, Liaoning Province, China.

Sinornithosaurus milleni

Chinese bird-lizard discovered near the turn of the millennium

The discovery of a near-complete skeleton of *Sinornithosaurus* was announced to the world in 1999 by Dr Xu Xing and colleagues as the first record of a dromaeosaurid being preserved with feathers. *Sinornithosaurus* was about one metre in length, with extremely long arms, the longest of any dinosaur relative to its leg length, and had peculiar pits and grooves on the depressed skull region in front of the eye. Its few teeth are much larger than those of other dromaeosaurids. A detailed study of its skull showed that it is has more features in common with the earliest bird, *Archaeopteryx*, than previously thought. There are at least three known specimens, representing two species, the most spectacular of which is nicknamed 'Dave', a juvenile animal with its long arms folded in a manner similar to birds' wings.

Artist's note: Here *Sinornithosaurus* is in a final stage of flight, with forewings drawn back and hind limbs brought forward in preparation for landing. Its weight-bearing toes are folded back and its second toe, with its enormous claw, is ready to take the weight of the animal. This claw may also have been of considerable use when shinnying up smooth-trunked trees. This is conjecture on my part, but it does introduce the possibility that the signature 'killer-claw' of these early dromaeosaurs had uses other than the butchering of prey.

Infraorder: Coelurosauria

Family: Dromaeosauridae

Age: Early Cretaceous

Locality: Sihetun region, Beipiao, Liaoning Province, China.

Saurornitholestes langstoni

The lizard-bird thief named in honour of Mr Langston

Saurornitholestes was the most abundant small theropod found in the late Cretaceous sediment of Dinosaur Provincial Park, Alberta, Canada. At about 1.6 metres in length, it was much smaller than its contemporary *Dromaeosaurus*. It also had distinctive teeth that were serrated quite differently from those of *Dromaeosaurus*. A tooth of *Saurornitholestes* found embedded in the wing bone of the giant pterosaur, *Quetzalcoatlus*, suggests that it fed upon the carcasses of this fallen giant. A second species, *Saurornitholestes robustus,* has been found from the San Juan Basin of New Mexico.

Artist's note: Here *Saurornitholestes* is feeding from the carcass of the gigantic pterosaur, *Quetzalcoatlus*. It is plausible that a coordinated attack by a pack of *Saurornitholestes* may have disabled a grounded pterosaur the size of *Quetzalcoatlus*, although it was more likely to have been scavenged upon. I have illustrated *Saurornitholestes* with blood staining to the naked areas of skin on the face and jaws. Sexual dimorphism is indicated by the purple colouring of the face, which contrasts with the yellow around the eye and gape of the mouth. White feathering of the malar and cheek areas complete this display.

Infraorder: Coelurosauria

Family: Dromaeosauridae

Age: Early Cretaceous

Locality: Dinosaur Provincial Park and other sites, Alberta, Canada; also Montana and New Mexico, USA.

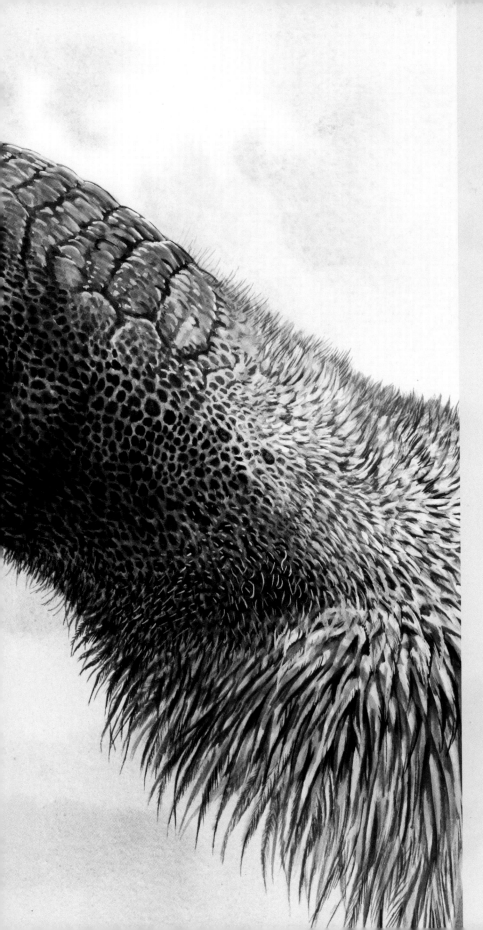

Tsaagan mangas
The white monster

The remains of *Tsaagan,* consisting of a well-preserved skull and eight neck vertebrae, were found in 1993 at the exotic-sounding location of Xanadu near Ukhaa Tolgod. *Tsaagan* was a contemporary of *Velociraptor,* but a much rarer animal in the desert ecosystem of ancient Mongolia. It was similar in size – around one-and-a-half metres in length – but had a slightly more robust skull and jaws. Because it differs from all other dromaeosaurids in a number of anatomical details of its skull and braincase, it is regarded as one of the most primitive members of the group. We have no direct evidence of what it ate or how it lived, but we can assume from its predatory teeth and powerful jaws that *Tsaagan* was an active predator that probably stalked the small mammals and medium-sized herbivorous dinosaurs of its habitat. Alternatively it might well have lived in small packs in order to tackle larger prey.

Artist's note: Reminiscent of a vulture, *Tsagaan* had a long, low skull and a narrow, upturned snout, which no doubt served it well when plunging into a carcass. Its teeth were of an efficient design, enabling it to tear flesh with ease.

Infraorder:	Coelurosauria
Family:	Dromaeosauridae
Subfamily:	Velociraptorinae
Age:	Late Cretaceous
Locality:	Djadokhta Formation, Ukhaa Tolgod, south-western Mongolia.

Dromaeosaurus albertensis
The running lizard from Alberta

This dinosaur, known from an incomplete skull and some associated bones of the feet and hand, was found in Alberta in 1914 by renowned dinosaur hunter Barnum Brown. *Dromaeosaurus* grew to about 1.8 metres in length. The skull is robust, lacking many of the hollow cavities that lighten the skulls of other members of the group. Similarly, the lower jaws are deeper than most dromaeosaurids, except for *Deinonychus*. The teeth of *Dromaeosaurus* are relatively large and strong, with those at the front of the mouth having a D-shaped cross-section. Perhaps its heavy teeth and skull structure made *Dromaeosaurus* a formidable hunter of large prey. There is no associated evidence that it hunted in a pack like *Deinonychus*.

Artist's note: Like others in this family, *Dromaeosaurus* has been illustrated without feathering on the face and lower jaw. Sexual dimorphism is indicated by the brightly coloured throat sac and the 'ruff' of orange and black feathers at the rear of the head. A body colour of black and brown stripes over white give this animal a cryptic advantage when stalking prey.

Infraorder: Coelurosauria

Family: Dromaeosauridae

Age: Late Cretaceous

Locality: Dinosaur Provincial Park, Alberta, Canada; Montana, USA.

Deinonychus antirrhopus
The terrible claw with a counter-balancing tail

The discovery of *Deinonychus* in 1931 near Bridger, Montana, by American dinosaur hunter Barnum Brown included a number of individuals together with the skeleton of a much larger herbivore, *Tenontosaurus*. This led Yale palaeontologist John Ostrom to infer that the animals hunted in a pack to take down the larger prey items, thus beginning the now popular idea that raptors were cunning pack hunters. Ostrom showed also that *Deinonychus* had light, hollow bones and other bird-like features, leading him to revive the idea that theropods were the closest relatives of birds. *Deinonychus* reached just over three metres in length, and weighed around 70 kilograms. It had particularly large slashing claws on the hand and feet relative to its body size. The bones of its tail had elongated bony processes that stiffened it for use as a counterbalance. Eggshell belonging to *Deinonychus,* preserved with the original finds, enabled the first dromaeosaurid eggs to be identified.

Artist's note: This reconstruction displays the incredible agility of these predators. The leader of the pack is the larger and stronger-coloured male on the right. He stands on two toes, using his feathered forearms and tail for balance. His hyper-extensile second toe, armed with a sickle claw, is deployed ready for action. With his other toes retracted, he draws the killer claw down with great force, slashing open his hapless victim. He targets the abdomen of the *Tenontosaurus,* an area that would probably have been free of the protective bony scutes and ossicles found on its back.

Infraorder: Coelurosauria

Family: Dromaeosauridae

Age: Early Cretaceous

Locality: Bridger, Montana, USA (Cloverly Formation).

Velociraptor mongoliensis
The swift thief from Mongolia

One of the most notorious dinosaurs of all time, *Velociraptor* was first discovered in the 1920s on an American expedition to the Gobi Desert. In 1971, a Polish–Mongolian expedition found a complete skeleton of *Velociraptor* locked in a battle with a sheep-sized *Protoceratops*, the result of a large dune of wet sand suddenly collapsing on the two dinosaurs and preserving them in life position. *Velociraptor* was about 1.8 metres in length and weighed close to 20 kilograms. Its hands bore three sharply re-curved claws, and its foot had an enlarged second toe claw which was retractable. Popular belief was that these claws could disembowel prey but tests have shown that they would not have been much use on the tough skin of large dinosaurs. Perhaps *Velociraptor* used these claws to scale the backs of large prey animals so they could use their deadly jaws to inflict wounds around the neck and spine. Eventually the prey animal would be weakened from the multiple attacks and succumb to the pack waiting for it.

Artist's note: The fossil find suggesting that this pair were entombed while locked in mortal combat has gripped popular imagination. It is equally plausible, however, that it reveals a scavenger at a carcass. Naturally, I have opted for the more dramatic interpretation, with the unfortunate *Protoceratops* being despatched by *Velociraptor* with an eviscerating kick to the abdomen.

Infraorder: Coelurosauria

Family: Dromaeosauridae

Age: Late Cretaceous

Locality: Djadokhta Formation, Omnogov, south-western Mongolia; also Inner Mongolia, China.

Bambiraptor feinbergi
The Bambi raptor named after the Feinberg family

The almost-complete skeleton of *Bambiraptor* was discovered by a 14-year-old Wes Linster near the Glacier National Park in Montana. It was hailed in the media as being 'The Rosetta Stone' of fossils for its value in understanding the bird-dinosaur transition. The small size of this little dromaeosaurid inspired palaeontologist David Burnham and his colleagues to name it after the Disney character Bambi. The species name is for the Feinberg family who bought the specimen and donated it to the Graves Museum in Florida. The specimen, currently housed at the American Museum of Natural History, is a juvenile that in life would have been about 70 centimetres in length, weighing around two kilograms. *Bambiraptor* had a very large brain, comparable in size to that of small birds. The ratio of its brain size to body weight is higher than for any other dinosaur. However, as it represents a juvenile, this may not hold in the adult where the brain is proportionately smaller. An interesting feature of *Bambiraptor* is that it had opposable front claws and the ability to hold prey to its mouth.

Artist's note: This animal is a late juvenile, evident from the large head and long limbs in relation to the body size. It has lost its down feathering and has begun to grow the contour feathers of the body and forelimbs which distinguish it as adult. A hint of adult colour pattern is apparent in the black and white markings of the face and forelimb feathers.

Infraorder: Coelurosauria

Family: Dromaeosauridae

Age: Late Cretaceous

Locality: Near Glacier National Park, Montana, USA.

Atrociraptor marshalli
The cruel thief named after Mr Marshall

This dinosaur is known from only one specimen comprising some parts of both jaws and teeth, found in the Drumheller region of Alberta, Canada. The jaws indicate that the skull was unusually deep with a short snout. The teeth of *Atrociraptor* are quite straight but emerge from the jaws angled backwards. They have large serrations along their cutting edges indicating that *Atrociraptor* liked to eat meat, sawing it off in large chunks. Despite its fearful appearance, *Atrociraptor* was a very small animal, reaching about 80 centimetres in length.

Artist's note: *Atrociraptor* was a most formidable predator. With a snout shorter and deeper than the other dromaeosaurs, it had jaws lined with a battery of strong, serrated teeth. The lack of feathering on the face and jaws is a common feature of flesh tearing, meat-eaters – evident in contemporary birds such as vultures. The targeted prey is the small hypsilophodont, *Parksosaurus warreni*.

Infraorder: Coelurosauria

Family: Dromaeosauridae

Age: Late Cretaceous

Locality: Drumheller, Alberta, Canada (Horseshoe Canyon Formation).

Jinfengopteryx elegans

The elegant golden Phoenix wing

A complete skeleton adorned with feathers was discovered in Hebei Province in central China in 2005. The specimen was at first described as being an early primitive bird. However, recent study shows that *Jinfengopteryx* had specialised features that are common to troodontids, such as an enlarged claw on its second toe, small nostrils, and a rounded snout profile. *Jinfengopteryx* was about 55 centimetres in length and had relatively long arms covered with branching feathers, as in birds, but lacked feathers around its legs. The gut region of the fossil shows small rounded structures – possibly seeds – indicating it may well have been a specialised plant eater or partially omnivorous. *Jinfengopteryx* demonstrates a clear linkage between the troodontids and the first birds.

Artist's note: The fossil upon which this reconstruction is based is of a female. The oval structures discovered in its abdominal region could alternatively have been undeveloped eggs. I have therefore furnished *Jinfengopteryx* with a coat of spots and bars that would afford it some camouflage, particularly when brooding a clutch of eggs amongst the leaf litter of the forest floor.

Infraorder: Coelurosauria

Family: Troodontidae

Age: Late Jurassic–Early Cretaceous

Locality: Hebei Province, China (Qiaotou Formation).

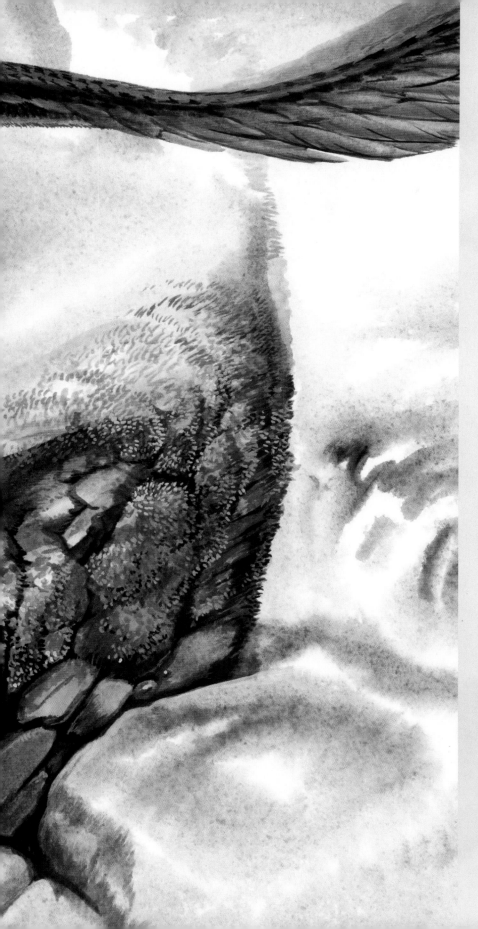

Sinovenator changae

Chinese hunter named in honour of Professor Meemann Chang

The almost complete skeleton of *Sinovenator* was found in Liaoning , north-eastern China in 2002. It was a small dinosaur, about a metre in length, and is regarded as one of the oldest and most primitive members of the troodontid family. It was named in honour of renowned Chinese palaeontologist Meemann Chang. The skeleton of *Sinovenator* showed that it was of slender build with long legs for fast running. Some features of its skeleton are found in birds and this indicates how closely related troodontids and primitive birds actually are. The braincase of *Sinovenator* closely resembles that of the first bird, *Archaeopteryx*.

Artist's note: Endowed with long, slender legs, *Sinovenator* was a fast little runner. Although it was incapable of flight, its feathered forelimbs may have helped it to make sudden directional changes during a chase and may also have been deployed for the entrapment of insect prey. In this reconstruction, *Sinovenator* has white markings around the eyes. similar to the owls of today. The single, contrasting, white dorsal stripe would help to disguise its shape when curled up at rest on the gloomy forest floor. This is a feature evident today among the *Dorcopsis* wallabies of the rainforests of New Guinea.

Infraorder: Coelurosauria

Family: Troodontidae

Age: Early Cretaceous

Locality: Yixian Formation, Lujiatun site near Beipiao city, Liaoning Province, China.

PETER SCHOUTEN · 2006

Byronosaurus jaffei
Byron Jaffe's lizard

This dinosaur was first discovered in 1993 and collected over several field seasons at Ukhaa Tolgod in Mongolia's Gobi Desert. It was named after Byron Jaffe, a sponsor of the American Museum's expeditions. *Byronosaurus* is represented by two individuals, one being a partial skeleton, the other comprising bits of a second skull. This dinosaur grew to about one-and-a-half metres in length and 50 centimetres in height, perhaps weighing about four kilograms. It had a distinctive elongated snout bearing numerous teeth. Its lower jaw showed a tight packing of its teeth at the front of its mouth. The skull of *Byronosaurus* is one of the best known in the troodontid group. It shows a peculiar feature – also seen in birds – where the air enters the nostrils into a chamber before passing through to the mouth.

Artist's note: The large eyes, pointed skull and tiny teeth are reminiscent of both the little bat-eared fox of Africa and the curious falanouc of Madagascar. Both have evolved numerous small teeth in response to a largely insectivorous diet. *Byronosaurus* may have used its small teeth at the front of its mouth for gripping prey, while the larger, more widely spaced teeth at the rear crushed the exoskeleton. With a well-developed visual acuity, it is plausible that visual signalling was an important aspect of the behaviour of the troodontids. For this reason, *Byronosaurus* has been illustrated with areas of brightly coloured skin on its face and wattles.

Infraorder: Coelurosauria

Family: Troodontidae

Age: Late Cretaceous

Locality: Djadokhta Formation, Ukhaa Tolgod, south-western Mongolia.

Mei long

Sleeping dragon

The almost complete skeleton of *Mei long* was discovered within volcanic ash beds in northern China in 2004. Only 53 centimetres in length, the creature was preserved in what appears to be a sleeping position with its head tucked neatly under its folded arm and its legs folded underneath its body. This is exactly the way modern birds and some mammals sleep to prevent heat loss from the head region, implying that this posture for sleeping evolved first in dinosaurs before birds. *Mei long* is distinguished from other troodontids by it extremely large nostrils and closely packed teeth in the upper jaw. An analysis of its skeletal features shows that *Mei* is placed at the base of the troodontid family, closely allied to *Sinovenator*.

Artist's note: This reconstruction shows a slumbering *Mei long* being slowly entombed by a gentle rain of volcanic ash. Why did it not attempt to flee? Perhaps it was already dead or, more plausibly, it may have adopted a 'freeze' position after being separated from its parent or presented with imminent danger. This behaviour is evident in birds such as plovers that habitually nest in the open. It allows the static and well-camouflaged young to be overlooked by a potential predator. The camouflage patterns that would ordinarily be visible on *Mei long* have been 'whited-out' by the ash.

Infraorder: Coelurosauria

Family: Troodontidae

Age: Early Cretaceous

Locality: Yixian Formation, Lujiatun site near Beipiao city, Liaoning Province, China.

Sinusonasus magnodens

Nose with sinusoidal shape, and having big teeth.

This troodontid is known from a near-complete single skeleton preserved on a slab of shale. Its teeth are large and robust and those at the front of the mouth lack serrations, making it distinct from other small troodontids. It was certainly an opportunistic predator that might have been capable of ambushing hard-to-catch prey such as pterosaurs resting on the ground or small mammals up in the trees. Like other troodontids it probably had well-developed feathers on its arms to speed its running through arm-flapping movements, or maybe leaping from the trees with a half-gliding motion to surprise prey. We do know that *Sinusonasus* lived amongst the very diverse assemblage of dinosaurs, birds, mammals, small reptiles and amphibians of the Jehol Biota in northern China. It had a wide variety of small prey items to choose from, but must have remained continuously wary of larger predators looking for an easy meal, which were also common at the time.

Artist's note: This recreation shows *Sinusonasus* hunting a pterosaur, in this instance the ornitho-cheirid, *Haopterus*. As with other troodontids, its feathered forelimbs played an important role in maintaining its stability when running at full flight. It is possible these forelimbs were also used in the capture of prey, bringing them forward and over the victim to prevent escape. Such use of the wings is commonplace among contemporary birds of prey.

Infraorder: Coelurosauria

Family: Troodontidae

Age: Early Cretaceous

Locality: Yixian Formation, Lujiatun site near Beipiao city, Liaoning Province, China.

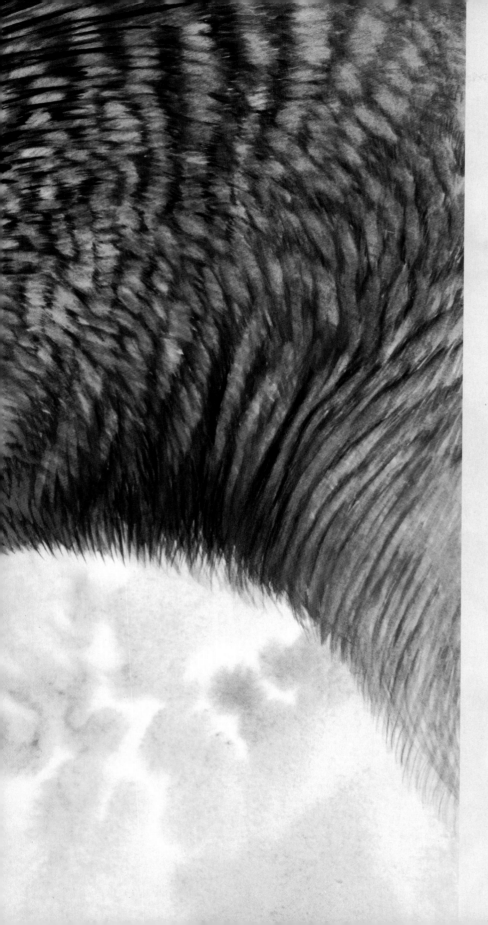

Saurornithoides mongoliensis

The lizard-bird form from Mongolia

This dinosaur was first discovered on the American Museum expedition to Mongolia in the 1920s, and described by renowned US palaeontologist Henry Fairfield Osborn in 1924. It was then known from a skull and partial body skeleton, and was the first troodontid to be described from Asia. Reaching two-and-a-half metres in length, *Saurornithoides* had large eyes, a narrow snout armed with many recurved teeth, and long grasping hands. Its eyes faced forwards giving it stereoscopic vision for accurate hunting. It has been estimated its brain had a mass around 100 grams, which – for an animal weighing about 35 kilograms – was extremely large compared with all other theropod dinosaurs. A second species, *Saurornithoides junior*, with a larger skull and longer snout, was found at Bugin Tsav in Mongolia and described by Rinchen Barsbold in 1974.

Artist's note: I have given *Saurornithoides* large, alert eyes, which are characteristic of the troodontids. The orientation of its eyes gives it overlapping fields of vision and some degree of binocularity. With such vision, these animals may have been well equipped for hunting in low light. The small teeth of *Saurornithoides* were not suitable for dealing with large prey. Its preferred diet probably consisted of the small mammals, amphibians and insects of its forested environment.

Infraorder: Coelurosauria

Family: Troodontidae

Age: Late Cretaceous

Locality: Djadokhta Formation, Ukhaa Tolgod, south-western Mongolia.

Sinornithoides youngi
Chinese bird form named after Dr Young

An almost complete skeleton, in a curled-up posture, was discovered in 1988 on the joint Canadian–Chinese expeditions to Inner Mongolia. *Sinornithoides* was the first relatively complete troodontid skeleton found, revealing the anatomical details of its arms, back and neck. It was just over a metre in length and weighed about eight kilograms. Its long tail comprised half its body length and retained a fair degree of flexibility. Its head was proportionately short compared to its body length, and its teeth were finely serrated along their rear edges. *Sinornithoides* was at first thought to be found in a sleeping posture like *Mei long*, but other experts dispute this and believe it was preserved huddling against an oncoming landslide or volcanic ash fall.

Artist's note: With a slender snout and triangular frontal bones, the skull of this animal indicated that it had a degree of binocular vision. With grasping hands and a switchblade claw on the second toe, *Sinornithoides* was no doubt an alert and agile little predator. On the basis that it was a nocturnal hunter, I have given it a black and white barred pattern like the night-hunting genets and civets of today. Its prey, about to be plucked from a gingko branch, is a small nocturnal mammal based on the multituberculate, *Sinobataar*.

Infraorder: Coelurosauria

Family: Troodontidae

Age: Early Cretaceous

Locality: Inner Mongolia, northern China.

Troodon formosus
Beautifully crafted wounding tooth

First described from the discovery of its teeth, most *Troodon* remains have been found as isolated bones, and nests of *Troodon* eggs have also been found at a site in Montana, USA. One nest included the partial remains of an adult, which led to the first suggestion of dinosaurs brooding their eggs. *Troodon* was around two metres in length and weighed up to 50 kilograms. Its had long arms that could fold back like a bird wing, forward-facing eyes for stereoscopic vision, and lower jaws that met in a U-shaped joint, similar to that of an iguana. The well-preserved braincase of *Troodon* shows it had an enlarged middle ear canal, giving it great acuity of hearing. This, combined with its large eyes, has led some scientists to suggest it could have been specially adapted for nocturnal hunting. *Troodon* may have favoured cooler climates as its teeth have been found as far north as the banks of the Coleville River, Alaska, then within the Cretaceous Arctic circle.

Artist's note: The cryptic colour and pattern on this brooding female blend it with its surrounds, even with its forearm feathers spread to protect its eggs and hatchlings from direct sunlight. I have given the newly emerged hatchling an egg tooth – a small horn on the top of its snout, used to break out of its egg. This egg tooth drops off once the young has hatched, so it is not surprising that there is no evidence for it in the fossil record. An egg tooth is a common feature of egg-laying animals not equipped with a beak, such as lizards or even the echidna.

Infraorder: Coelurosauria

Family: Troodontidae

Age: Late Cretaceous

Locality: Drumheller, Alberta, Canada; also Montana and Alaska, USA.

Yandangornis longicaudus
Yandang bird with a long tail

Yandangornis is another primitive bird-like form found from Zhejiang Province in south-eastern China. It was first described as an early fossil bird represented by a relatively complete skeleton. However, Dr Zhonghe Zhou makes the point that it actually has no definite feature making it a bird, so we are inclined to regard it as another small winged predatory dinosaur, perhaps a troodontid. It exemplifies the difficulty in drawing the line between actual birds and highly evolved feathered dinosaurs, even by experts who spend their lives studying such fossils. It was about 50 centimetres in length.

Artist's note: This remarkably bird-like dinosaur was well adapted for life on the forest floor. I have based its colour and cryptic pattern on the forest-dwelling tinamous of South America, and its pointed, hair-like feathers on the flightless kiwis of New Zealand. Their softened appearance is the result of the loss of barbules from the barbs of the feathers, which prevents them from being able to 'zip' together into the stiffened plane of a normal plume. This type of plumage provides good insulation and it is also very effective at shedding rain.

Infraorder: Coelurosauria

Family: Indeterminate

Age: Late Cretaceous

Locality:, Zhejiang Province, south-eastern China.

Epidendrosaurus ningchengensis

The upon tree lizard from Ningcheng

Epidendrosaurus was the first non-avian dinosaur found to have adaptations to climbing trees, as evidenced by its long hands and strongly curved foot claws. The only specimen known is probably a juvenile animal, roughly the size of a sparrow. *Epidendrosaurus* was equipped with an elongated third finger, similar to that of the modern aye-aye, which uses it finger to extract grubs from within the trunks of trees. A very similar dinosaur, *Scansoriopteryx,* from Liaoning Province, China, is thought by experts to be the same genus as *Epidendrosaurus.*

Artist's note: In this illustration, a parent is bringing food to its young. Its nest, situated in the crown of a tree fern, has been constructed of mosses, lichens and plant fibres. The nest has been inspired by those of the birds of paradise, many of which locate their nests on the crown of a tree fern. They do this because tree ferns often emerge from the surrounding vegetation, and, with their solitary trunk and fronds covered with irritating scales, make access for predators difficult.

Infraorder: Coelurosauria

Family: Scansoriopterygidae

Age: Middle Jurassic–Early Cretaceous (unresolved), but now thought closer to basal Cretaceous Jehol Group.

Locality: Near Ningcheng (Daohugou Beds) Inner Mongolia, north-eastern China.

Shuvuuia deserti
The desert bird

Shuvuuia is the most complete member of the enigmatic alvarezsaurid family. It was a small animal, about 60 centimetres in length, with a gracile frame, long powerful legs, and a narrow skull with elongated jaws bearing small teeth. Like other members of its group it had short, powerful arms with one very large claw and two much smaller claws. A fascinating discovery was that *Shuvuuia* had feathers, as the original skeleton had thin tube-like structures surrounding it. On closer analysis these turned out to be decayed structures made of certain proteins similar to that of modern bird feathers. *Shuvuuia* also had the ability to flex its upper jaw independently of the braincase, a feature of modern birds not seen in any other dinosaur. It is no surprise that the remains of *Shuvuuia* were first classified as being a very primitive flightless bird rather than a dinosaur.

Artist's note: Illustrating the bizarre and enigmatic alvarezsaurs presents a great challenge. What type of lifestyle did they lead? Perhaps *Shuvuuia* was a specialist feeder, with a diet of insects and other invertebrates. It may have used its powerful, stubby arms with their massive claw for smashing open termite mounds, chiselling into rotten logs or prizing behind bark in the search for prey. As protection from the biting insects, I have given *Shuvuuia* dense, feathers around its head and neck. The long, well-feathered tail provides a counter-balance when hammering open a rotten log, and helps to insulate and camouflage the animal when curled up at rest.

Infraorder: Coelurosauria

Family: Alvarezsauridae

Age: Late Cretaceous

Locality: Ukhaa Tolgod and Togrogin Shiree, Mongolia.

Mononykus olecranus

The single claw with well-developed olecranon

The discovery of this dinosaur was a joint collaboration between Mongolian and US scientists who both had discovered remains of the same dinosaur from the same region on different expeditions. In 1992 Mongolian palaeontologist Perle Altangarel visited the Museum of Natural History in New York to show the strange bones of a turkey-sized dinosaur he had found to US palaeontologists Luis Chiappe, Mark Norell and James Clark. They too had similar bones from the Gobi desert sites. They named the animal *Mononykus* as it had just one large claw on the forelimb, and a well-developed olecranon or 'funny bone'. *Mononykus* was about one metre in length, and probably ate insects.

Artist's note: This animal has been reconstructed on the supposition that it fed on insects and beetle larvae. It has long pennaceous tail feathers and a dense feathering to the neck. Stark contrasts of black and white help to break up the outline of its body and the red flush to the face could be a signalling feature that becomes brighter and more widespread among courting males.

Infraorder:	Coelurosauria
Family:	Alvarezsauridae
Age:	Late Cretaceous
Locality:	Bugin Tsav, Mongolia.

Archaeopteryx lithographica
The ancient wing from the lithographic shales

The most famous of all early bird fossils, *Archaeopteryx* was first named from a single feather found in 1860 at Eichstätt in southern Germany. Since that time, some 10 specimens of *Archaeopteryx* have been found and studied in minute detail. We know it was a true bird with complex feather patterns, capable of powered flight, and that it could probably take off from a standing position. It had small teeth and a long tail with feathers fanning outwards. Its arms display asymmetrical flight feathers as occurs in modern birds. Detailed CT scans of its braincase confirm that, like modern birds, it had similar sensory developments to process information rapidly during flight. *Archaeopteryx* grew to about 70 centimetres in length with a 50 centimetres wingspan, but weighed less than one kilogram. It narrow, toothed beak suggests its was primarily an insect feeder.

Artist's note: A good contemporary analogy for *Archaeopteryx* are rails, particularly those found on arid islands. Of equivalent size and capable of flight, rails are suited more to a life on the ground than in trees. Like these opportunistic birds, *Archaeopteryx* may have roamed the vegetation bordering watercourses or marshland in search of prey, taking to wing only when danger threatened. Inspired by the uniformly conservative appearance of the rails, *Archaeopteryx* has been endowed with a colour scheme of drab browns and grey, with some faint cryptic barring.

Order: Archaeopterygiformes

Family: Archaeopterygidae

Age: Late Jurassic

Locality: Southern Germany (Solnhofen shales).

Jeholornis prima (*Shenzhouraptor*)

The first Jehol bird

Jeholornis was one of the largest known Cretaceous birds, second only to *Sapeornis*, being just under a metre in length. It had a long reptilian tail like *Archaeopteryx* but had advanced features in its hand that indicate that it had powerful flight capabilities. *Jeholornis* was found with numerous rounded structures inside its gut, thought to be seeds from its last meal. The description of a closely related form called *Shenzhouraptor*, published at about the same time as *Jeholornis*, has caused a degree of confusion in the literature as some scientists believe these to be the same animal. The advice of the authors, Zhonghe Zhou and Fucheng Zhang, is that the name *Jeholornis* was actually published first and thus takes priority. *Jeholornis* demonstrates that adaptions to flying evolved first in the wing of birds before the tail was modified to a shorter length.

Artist's note: In order to metabolise sufficient protein and energy from its probable diet of seeds or fruit, *Jeholornis* would need to expend considerable time and energy in digestion. I have illustrated it perching with its outstretched wings and back to the sun in order to help replenish its energy. The dark-coloured feathers act as a solar collector by absorbing heat, thereby maintaining body temperature and diminishing the need for *Jeholornis* to expend its own energy for this purpose.

Order: Jeholornithiformes

Family: Jeholornithidae

Age: Early Cretaceous

Locality: Chaoyang city region, Liaoning Province, China (Yixian Formation).

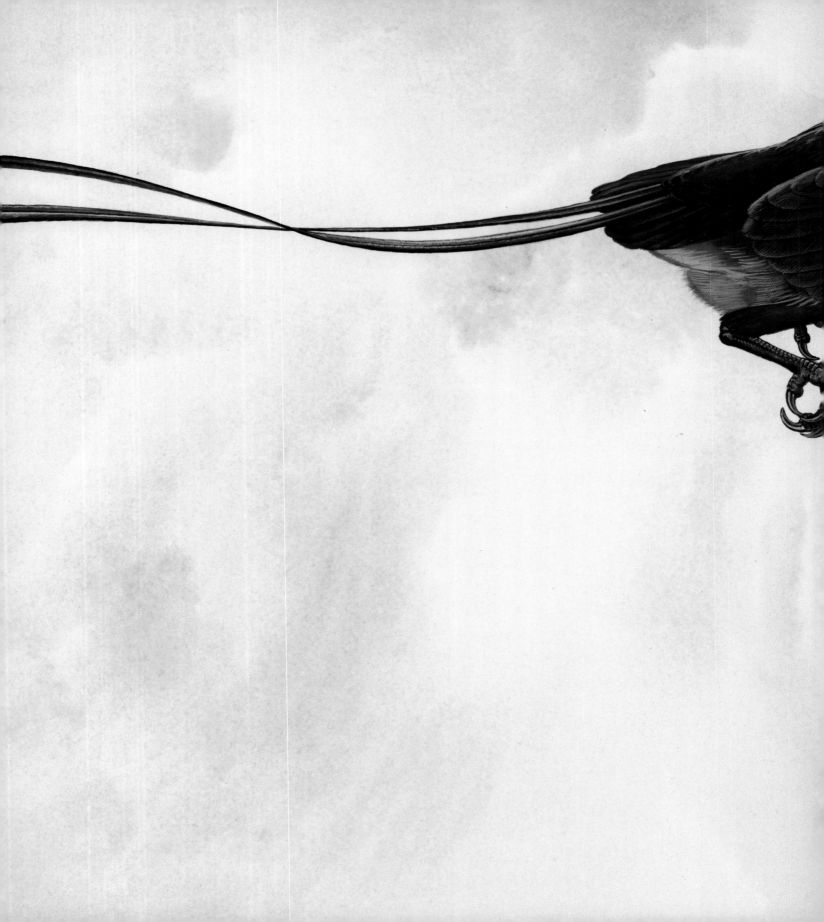

Confuciusornis sanctus
The sacred Confucius bird

Confuciusornis is one of the best known of all Mesozoic age birds, with over a thousand good specimens of complete, feathered individuals in Chinese museums. About the size of a modern crow, *Confuciusornis* differed principally from *Archaeopteryx* in having a toothless beak and lacking the latter's long reptilian tail. It also had a pygostyle, or shortened tail bone, a feature of modern birds that helps them to manoeuvre the tail in flight separately from the hind limbs. Its wrist shows partial fusion, a condition more specialised than occurs in *Archaeopteryx*, thus enabling a stronger wing design. The many complete specimens of *Confuciusornis* show some with long tail feathers and others without them, suggesting this feature was a difference exhibited between males and females. A recent review of Chinese fossil birds has revealed that *Confuciusornis* appeared to have leg feathers.

Artist's note: With a short body, well-developed wings and a reduced tail, this was an animal that spent substantial time aloft. The large number of individuals found in fossil deposits indicate that *Confuciusornis* formed large flocks. With its aeronautical skills, it is possible that *Confuciusornis* could display whilst on the wing. The bright blue and reflective quality of the feathers have been influenced by the metallic starlings of Africa. These birds, as a flock and with their brilliant colouring, are able to confuse potential predators by presenting such a united, dazzling display that it becomes very difficult to isolate an individual.

Family: Confuciusornithidae

Age: Early Cretaceous

Locality: Sihetun sites, near Beipiao city, Liaoning Province, China (Chaomidianzi Formation).

Changchengornis hengdaoziensis

Changcheng bird from the Hengdaozi region

Changchengornis was the second of the *Confuciusornis*-like birds to be recognised from the many hundreds of specimens from the small quarry near Jiangshangou village in Liaoning, China, yet it is still only known from one specimen. It is of similar size but differs from *Confuciusornis* in having a shorter, more strongly curved beak and in minor details of the body skeleton, such as shape of the wishbone, and proportions of the wrist bones. It also had a long tail formed of two large tail feathers, possibly as long and the body. The foot structure of *Changchengornis* indicates it had slightly more powerful grasping ability than *Confusciusornis*.

Artist's note: This reconstruction has been inspired by birds of paradise and cotingas, both of which display in the shady world under the forest canopy where bright sunlight is at a premium. The scattered areas where the sun pierces through provide prime display sites for rainforest birds. Common features of many of these species are areas of white, blue or refractive feathering, designed to be reflective of the ultraviolet spectrum of sunlight. With the open-mouthed gape of the male *Changchengornis*, there is the suggestion that vocalisations were a part of its courtship repertoire.

Family: Confuciusornithidae

Age: Early Cretaceous

Locality: Sihetun sites near, Beipiao city, Liaoning Province, China (Chaomidianzi Formation).

Sapeornis chaoyangensis

Bird honouring the Society for Avian Paleontology and Evolution from Chaoyang County

Known from about half-a-dozen well-preserved specimens from Liaoning, China, *Sapeornis* was the largest known Cretaceous bird. It had a wingspan of almost a metre – around twice the size of *Archaeopteryx*. Its skull was armed with protruding teeth in the upper jaw. Despite its large wings, it lacked a well-developed sternum and thus had not yet evolved the modern style of flight that characterises later birds. The fossils also show the remains of small grinding stones in its stomach, suggesting it probably ate hard food items such as seeds. *Sapeornis* is thought to be a bird at a similar stage of evolution as *Confuciusornis,* but demonstrates that some birds had at least evolved an effective mode of flight that would work with larger body sizes.

Artist's note: With its elongated forelimbs, short hind limbs and a reduced tail ending in a pygostyle, *Sapeornis* must surely have been a most adept flier. With an overall pattern designed to be cryptic, ornament has been restricted to the bright yellow commissure of the mouth and the bright red throat sac. The darker colour of the flight feathers indicates a higher level of melanin. This helps to strengthen the feathers as melanin granules are always associated with increased amounts of keratin – a structural component of hair, feathers and claws.

Order: Sapeornithiformes

Family: Sapeornithidae

Age: Early Cretaceous

Locality: Jiufotang Formation, Chaoyang County, Liaoning Province, China.

Jibeinia luanhera

One from Ji Bei near the Luanhe River

Jibeinia was a sparrow-sized creature, about 12 centimetres in length, with a few small, pointed teeth in its beak. The body skeleton shows advanced features for flying such as a strut-like shoulder bone (coracoid) and a pygostyle supporting the two long tail feathers. *Jibeinia* is seen to be more like modern birds than *Sapeornis* or *Confuciusornis* in many features, noticeably in its V-shaped wishbone, having reduced claws on the hands, and only two finger bones present on the third finger. It is now regarded as an early member of the Enantiornithines, a group located at the beginning of the modern bird radiation. It was an early flying bird that resembled in general form a typical flying bird of today, except perhaps in the visible signs of its external finger bones. Once birds like *Jibeinia* had taken to the skies the world would never be the same place again.

Artist's note: Like *Confuciusornis* and *Chang-chengornis*, this animal also had a pair of long tail feathers. The presence of teeth, however, distinguishes *Jibeinia* from these species. The presence of the tail feathers may indicate a male, although this is speculative as *Jibeinia* is known to us by only one specimen. This reconstruction is of a male, sporting bold colours and a reflective throat shield reminiscent of the smaller birds of paradise.

Order: Protopterygiformes

Family: Protopterygidae

Age: Early Cretaceous

Locality: Near Tuyao village, Fengning County, northern Hebei Province, China (Yixian Formation).

SELECTED REFERENCES

Alonso, P.D., Milner, A.C., Ketcham, R.A., Cookson, M.J. & Rowe, T.B. 2004. The avian nature of the brain and inner ear of Archaeopteryx. *Nature* **430**: 666–669.

Baier, D.B, Gatesy, S.M. & Jenkins Jr, F. 2007. A critical ligamentous mechanism in the evolution of avian flight. *Nature* **445**: 307–310.

Barsbold, R. 1974. Saurornithoididae, a new family of theropod dinosaurs from Central Asia and North America. *Paleontologica Polonica* **30**: 5–22.

Barsbold, R. 1983. Carnivorous dinosaurs from the Cretaceous of Mongolia. *Transactions of the Joint Soviet-Mongolian Paleontological Expedition* **19**: 5–119.

Carpenter, K. (ed.) 2005. *The Carnivorous Dinosaurs*. Indiana University Press, Bloomington, USA. 371pp.

Carr, T.D., Williamson, T.E. & Schwimmer, D.R. 2005. A new genus and species of tyrannosauroid from the Late Cretaceous (Middle Campanian) Demopolis Formation of Alabama. *Journal of Vertebrate Paleontology* **25**: 119–143.

Chang, M.M, Chen. P.-J., Wang, Y.Q., Wang, Y. & Miao, D. 2003. *The Jehol Biota. The emergence of feathered dinosaurs, beaked birds and flowering plants*. Shanghai Scientific and Technical Publications, Shanghai. 208pp.

Chiappe, L.M., Ji, S., Ji, Q. & Norell, M.A. 1999. Anatomy and systematics of the Confuciusornithidae (Theropoda: Aves) from the late Mesozoic of northeastern China. *Bulletin of the American Museum of Natural History* **242**: 1–89.

Chiappe, L.M. 2007. *Glorified Dinosaurs. The Origin and Early Evolution of Birds*. Wiley & Sons, USA; UNSW Press, Sydney, 263pp.

Chiappe, L., Norell, M.A. & Clark, J.M. 1996. Phylogenetic position of Mononykus from the Late Cretaceous of the Gobi Desert. *Memoirs of the Queensland Museum* **39**: 557–582.

Chen, P. Dong, Z. & Zhen, S. 1998. An exceptionally well-preserved theropod dinosaur from the Yixian Formation of China. *Nature* **391**: 147–152.

Clark, J.M., Norell, M.A. & Chiappe, L.M. 199). An oviraptorid skeleton from the Late Cretaceous of Ukhaa Tolgod, Mongolia, preserved in an avianlike brooding position over an oviraptorid nest.' *American Museum Novitates* **3265**: 36.

Clark, J.M., Norell, M.A. & Barsbold, R. 2001. Two new oviraptorids (Theropoda:Oviraptorosauria), upper Cretaceous Djadokhta Formation, Ukhaa Tolgod, Mongolia. *Journal of Vertebrate Paleontology* **21**: 209–213.

Currie, P.J. 1995. New information on the anatomy and relationships of *Dromaeosaurus albertensis*. *Journal of Vertebrate Paleontology* **7**: 72–81.

Currie, P.J. 2001. Anatomy of *Sinosauropteryx prima* from Liaoning, northeastern China. *Canadian Journal of Earth Sciences* **38**: 1705–1727.

Currie, P.J., Hurum, J.H. & Sabath, K. 2003. Skull structure and evolution in tyrannosaurid phylogeny. *Acta Palaeontologica Polonica* **48**: 227–234.

Erickson, G.M., Makovicky, P.J., Currie, P.J., Norell, M.A., Yerby, S.A. & Brochu, C.A. 2004. Gigantism and comparative life-history parameters of tyrannosaurid dinosaurs. *Nature* **430**: 772–775.

Farlow, J.O., Smith, M.B. & Robinson, J.M. 1995. Body mass, bone 'strength indicator', and cursorial potential of *Tyrannosaurus rex*. *Journal of Vertebrate Paleontology* **15**: 713–725.

Feduccia, A. & Tordoff, H. B. 1979. Feathers of Archaeopteryx: asymmetric vanes indicate aerodynamic function. *Science* **203**: 1021–1022.

Horner, J. R. & Lessem, D.1993. *The Complete T. Rex/How Stunning New Discoveries Are Changing Our Understanding of the World's Most Famous Dinosaur*. Simon and Schuster.

Hurum J.H. & Sabath, K. 2003. Giant theropod dinosaurs from Asia and North America: Skulls of *Tarbosaurus bataar* and *Tyrannosaurus rex* compared. *Acta Palaeontologica* Polonica **48**: 161–190.

Hutt, S., Naish, D., Martill, D.M., Barker, M.J. & Newbery, P. 2001. A preliminary account of a new tyrannosauroid theropod from the Wessex Formation (Cretaceous) of southern England. *Cretaceous Research* **22**: 227–242.

Huxley, T.H. 1867. On the classification of birds and on the taxonomic value of the modifications of certain of the cranial bones observable in that class. *Proceedings of the Zoological Society of London 1867*: 415–472.

Huxley, T.H. 1868. On the animals which are most nearly intermediate between birds and reptiles. *Annals and Magazine of Natural History, London* **2**: 66–75.

Hwang, S.H. 2004. A large compsognathid from the Early Cretaceous Yixian Formation of China'. *Journal of Systematic Paleontology* **2**: 13–39

Hwang, S.H., Norrell, M.A., Ji, Q. & Gao, K.Q. 2002. New specimens of *Microraptor zhaoianus* (Theropoda: Dromaeosauridae) from northeastern China. *American Museum Novitates* **3381**: 1–44.

Ji, S., Ji, Q., Lu J. & Yuan, C. 2007. A new giant compsognathid dinosaur with long filamentous integuments from Lower Cretaceous of northeastern China. *Acta Geologica Sinica* **81**: 8–15.

Kirkland, J.I., Zanno, L.E., Sampson, S.D., Clark, J.M. & DeBlieux, D.D. 2005. A primitive therizinosauroid dinosaur from the Early Cretaceous of Utah. *Nature* **435**: 84–87.

Kobayashi, Y., Lü, J.–C., Dong, Z.–M., Barsbold, R., Azuma, Y. & Tomida, Y. 1999. Herbivorous diet in an ornithomimid dinosaur. *Nature* **402**: 480–481.

Kobayashi, Y. & Lü, J.–C. 2003. A new ornithomimid dinosaur with gregarious habits from the Late Cretaceous of China. *Acta Palaeontologica Polonica* **48**: 235–259.

Lambe, L.M. 1914. On a new genus and species of carnivorous dinosaur from the Belly River Formation of Alberta, with a description of the skull of Stephanosaurus marginatus from the same horizon. *Ottawa Naturalist* **28**: 13–20.

Lü, J. 2002. A new oviraptorosaurid (Theropoda: Oviraptorosauria) from the Late Cretaceous of southern China. *Journal of Vertebrate Paleontology* **22**: 871–875.

Lü, J. 2005. *Oviraptorid dinosaurs from Southern China*. Beijing: Geological Publishing House. 200pp.

Maleev, E.A. 1955. [Gigantic carnivorous dinosaurs of Mongolia]. *Doklady Akademii Nauk S.S.S.R.* **104**(4): 634–637. [In Russian]

Makovicky, P.J., Apesteguia, S. & Agnolin, F.L. 2005. The earliest dromaeosaurid therpod from South America. *Nature* **437**: 1007–1010.

Martin, L. 1983. The origin of birds and of avian flight. *Current Ornithology* **1**: 105–129.

Maxwell, W.D. & Ostrom, J.H. 1995. Taphonomy and paleobiological implications of Tenontosaurus-Deinonychus associations. *Journal of Vertebrate Paleontology.* **15**: 707–712.

Nicholls, E. L. & Russell, D. 1985. The structure and function of the pectoral girdle and forelimb of *Struthiomimus altus* (Theropoda: ornithomimidae) *Palaeontology* **28**: 638–677.

Norell, M.A., Clark, J.M., Dashzeveg, D., Barsbold, T., Chiappe, L.M., Davidson, A.R., McKenna, M.C. & Novacek, M.J. 1994. A theropod dinosaur embryo, and the affinities of the Flaming Cliffs Dinosaur eggs. *Science* **266**: 779–782.

Norrell, M.A., Clark, J.M., Chiappe, L.M. & Dashzeveg, D. 1995. A nesting dinosaur. *Nature* **378**: 774–776.

Norell, M.A., Clark, J.M., Turner, A.H., Makovicky, P.J., Barsbold, R. & Rowe, T. 2006. A new dromaeosaurid theropod from Ukhaa Tolgod (Omnogov, Mongolia). *American Museum Novitates* **3545**: 1–51.

Norell, M.A. & Makovicky, P.J. 1999. Important features of the dromaeosaurid skeleton II: information from newly collected specimens of *Velociraptor mongoliensis*. *American Museum Novitates* **3282**: 1–45.

Novas, F.E. & Puerta, P.F. 1997. New evidence concerning avian origins from the Late Cretaceous of Patagonia. *Nature* **387**: 390–392.

O'Connor, P.M. & Claessens, L.P.A.M. 2005. Basic avian pulmonary design and flow-through ventilation in non-avian theropod dinosaurs. *Nature* **436**: 253–256.

Osborn, H.F. 1906. Tyrannosaurus, Upper Cretaceous carnivorous dinosaur (second communication). *Bulletin of the American Museum of Natural History* **22**: 281–296.

Osborn, H.F. 1917. Skeletal adaptations of Ornitholestes, Struthiomimus, Tyrannosaurus. *Bulletin of the American Museum of Natural History* **35**: 733–71.

Osborn, H.F. 1924a. Three new Theropoda, Protoceratops zone, central Mongolia. *American Museum Novitates* **144**: 1–12.

Osmólska H, Roniewicz E & Barsbold R. 1972. A new dinosaur, *Gallimimus bullatus* n. gen.,n. sp. (Ornithomimidae) from the Upper Cretaceous of Mongolia. *Paleontologica Polonica* **27**: 103–143.

Ostrom, J.H. 1969. Osteology of *Deininychus antirrhopus*, an unusual theropod from the Lower cretaceous of Montana. *Bulletin of the Peabody Museum of Natural History* **30**: 1–169.

Owen, R. 1862. On the fossil remains of a long-tailed bird (*Archaeopteryx macrurus*, Ow.) from the Lithographic Slate of Solenhofn. *Proceedings of the Royal Society of London* **12**: 272–273.

Padian, K. (Ed.). 1986. The origin of birds and the evolution of flight. *Memoirs of the Californian Academy of Sciences* **8**: 1–98.

Parker, W.K. 1864. Remarks on the skeleton of Archaeopteryx; and on the relations of the bird to the reptile. *Geological Magazine* **1**: 55–57.

Perez-Moreno, B.P., Sanz, J.L., Buscalioni, A.D., Moratalla, J.J., Ortega, F. & Raskin-Gutman, D. 1994. Unique multitoothed ornithomimosaur from the Lower Cretaceous of Spain. *Nature* **30**: 363–367.

Perle, A. 1977. [On the first finding of Alectrosaurus (Tyrannosauridae, Theropoda) in the Late Cretaceous of Mongolia.] *Problems of Geology of Mongolia* **3**: 104–113 [in Russian].

Perle, A. 1981. [A new segnosaurid from the Upper Cretaceous of Mongolia.] *Trudy Sovm. Soviet-Mongolian Ekspeditions* **15**: 50–59 [in Russian].

Perle, A., Norrell, M.A., Chiappe, L.M. & Clark, J.M. 1993. Flightless bird from the Cretaceous of Mongolia. *Nature* **362**: 623–626.

Russell, D.A. 1970. Tyrannosaurs from the Late Cretaceous of western Canada. *National Museum of Natural Sciences Publications in Paleontology* **1**: 1–34.

Russell, D. 1972. Ostrich dinosaurs from the Late Cretaceous of Western Canada. *Canadian Journal of Earth Sciences* **9**: 375–402.

Russell, D.A. & Dong Z. 1993. The affinities of a new theropod from the Alxa Desert, Inner Mongolia, People's Republic of China. In: Currie, P.J. (Ed.). Results from the Sino-Canadian Dinosaur Project. *Canadian Journal of Earth Sciences* **30**: 2107–2127.

Schweitzer, M.H., Wittmeyer, J.L., Horner, J.R. & Toporski, J.B. 2005. Soft tissue vessels and cellular preservation in *Tyrannosaurus rex*. *Science* **307**: 1952–1955.

Wagner, A. 1861. Neue Beiträge zur Kenntnis der urweltlichen Fauna des lithographischen Schiefers; V. Compsognathus longipes Wagner. *Abhandlungen der Bayerischen Akademie der Wissenschaften* **9**: 30–38.

Weishampel, D.B., Dodson, P. & Osmolska, H. (Eds) 2004. *The Dinosauria*. 2nd edn. University of California Press, Berkeley and Los Angeles, CA. 861pp.

Xu X., Clark, J.M., Forster, C.A., Norell, M.A., Erickson, G.M., Eberth, D.A., Jia, C. & Zhao, Q. 2006. A basal tyrannosauroid dinosaur from the Late Jurassic of China. *Nature* **439**: 715–718.

Xu, X. & Norell, M.A. 2004. A new troodontid dinosaur from China with avian-like sleeping posture. *Nature* **431**: 838–841.

Xu, X., Norell, M. A., Kuang, X., Wang, X., Zhao, Q. & Jia, C. 2004. Basal tyrannosauroids from China and evidence for protofeathers in tyrannosauroids. *Nature* **431**: 680–684.

Xu, X., Tang, Z-L. & Wang, X-L. 1999. A therizinosauroid dinosaur with integumentary structures from China. *Nature* **399**: 350–354.

Xu. X., Zhou, Z. & Prum, R.O. 2001. Branched integumental structures in Sinornithosaurus and the origin of feathers. *Nature* **410**: 200–203.

Xu. X., Zhou, Z., Wang, X., Kuang, F., Zhang, F. & Du, X. 2003. Four-winged dinosaurs from China. *Nature* **421**: 335–340.

Zhang, F. & Zhou, Z. 2004. Leg feathers in an Early Cretaceous bird. *Nature* **431**: 925.

Zhou, Z. & Zhang, F. 2002. A long-tailed seed-eating bird from the Early Cretaceous of China. *Nature* **418**: 405–409.

Zhou, Z. & Zhang, F. 2003. Anatomy of the primitive bird *Sapeornis chaoyangensis* from the Early Cretaceous of Liaoning, China. *Canadian Journal of Earth Sciences* **40**: 731–747.

INDEX